SpringerBriefs in Mathematical Physics

Volume 13

SpringerBriefs are characterized in general by their size (50-125 pages) and fast production time (2-3 months compared to 6 months for a monograph).
Briefs are available in print but are intended as a primarily electronic publication to be included in Springer's e-book package.

Typical works might include:

An extended survey of a field
A link between new research papers published in journal articles
A presentation of core concepts that doctoral students must understand in order to make independent contributions
Lecture notes making a specialist topic accessible for non-specialist readers.

SpringerBriefs in Mathematical Physics showcase, in a compact format, topics of current relevance in the field of mathematical physics. Published titles will encompass all areas of theoretical and mathematical physics. This series is intended for mathematicians, physicists, and other scientists, as well as doctoral students in related areas.

More information about this series at http://www.springer.com/series/11953

J.-B. Bru · W. de Siqueira Pedra

Lieb-Robinson Bounds for Multi-Commutators and Applications to Response Theory

 Springer

J.-B. Bru
Departamento de Matemáticas, Facultad de
 Ciencia y Tecnología
Universidad del País Vasco
Bilbao
Spain

W. de Siqueira Pedra
Department of Mathematical Physics,
 Institute of Physics
University of São Paulo
São Paulo
Brazil

and

IKERBASQUE, Basque Foundation
 for Science
Bilbao
Spain

and

BCAM—Basque Center for Applied
 Mathematics
Bilbao
Spain

ISSN 2197-1757   ISSN 2197-1765   (electronic)
SpringerBriefs in Mathematical Physics
ISBN 978-3-319-45783-3   ISBN 978-3-319-45784-0   (eBook)
DOI 10.1007/978-3-319-45784-0

Library of Congress Control Number: 2016950360

Mathematics Subject Classification (2010): 82C10, 82C20, 82C22, 47D06, 58D25, 82C70, 82C44, 34G10

Printed on acid-free paper

This Springer imprint is published by Springer Nature
The registered company is Springer International Publishing AG
The registered company address is: Gewerbestrasse 11, 6330 Cham, Switzerland

Contents

Abstract

We generalize to multi-commutators the usual Lieb–Robinson bounds for commutators. In the spirit of constructive QFT, this is done so as to allow the use of combinatorics of minimally connected graphs (tree expansions) in order to estimate time-dependent multi-commutators for interacting fermions. Lieb–Robinson bounds for multi-commutators are effective mathematical tools to handle analytic aspects of the dynamics of quantum particles with interactions which are non-vanishing in the whole space and possibly time-dependent. To illustrate this, we prove that the bounds for multi-commutators of order three yield existence of fundamental solutions for the corresponding non-autonomous initial value problems for observables of interacting fermions on lattices. We further show how bounds for multi-commutators of an order higher than two can be used to study linear and non-linear responses of interacting fermions to external perturbations. All results also apply to quantum spin systems, with obvious modifications. However, we only explain the fermionic case in detail, in view of applications to microscopic quantum theory of electrical conduction discussed here and because this case is technically more involved.

Chapter 1
Introduction

Lieb–Robinson bounds are upper-bounds on time-dependent commutators and were originally used to estimate propagation velocities of information in quantum spin systems. They have first been derived in 1972 by Lieb and Robinson [LR]. Nowadays, they are widely used in quantum information and condensed matter physics. Phenomenological consequences of Lieb–Robinson bounds have been experimentally observed in recent years, see [Ch].

For the reader's convenience and completeness, we start by deriving such bounds for fermions on the lattice with (possibly non-autonomous) interactions. As explained in [NS] in the context of quantum spin systems, Lieb–Robinson bounds are only expected to hold true for systems with short-range interactions. We thus define Banach spaces $\mathcal{W}$ of short-range interactions and prove Lieb–Robinson bounds for the corresponding fermion systems. The spaces $\mathcal{W}$ include density–density interactions resulting from the second quantization of two-body interactions defined via a real-valued and integrable interaction kernel $v(r) : [0, \infty) \to \mathbb{R}$. Considering fermions with spin $1/2$, our setting includes, for instance, the celebrated Hubbard model (and any other system with finite-range interactions) or models with Yukawa-type potentials. Two-body interactions decaying polynomially fast in space with sufficiently large degree are also allowed, but the Coulomb potential is excluded because it is not summable at large distances. The method of proof we use to get Lieb–Robinson bounds for non-autonomous C^*-dynamical systems related to lattice fermions is, up to simple adaptations, the one used in [NS] for (autonomous) quantum spin systems. Compare Theorem 4.3, Lemma 4.4, Theorem 5.1 and Corollary 5.2 with [NS, Theorems 2.3. and 3.1.]. See also [BMNS] where (usual) Lieb–Robinson bounds for non-autonomous quantum spin systems have already been derived [BMNS, Theorems 4.6].

© The Author(s) 2017
J.-B. Bru and W. de Siqueira Pedra, *Lieb-Robinson Bounds for Multi-commutators and Applications to Response Theory*, SpringerBriefs in Mathematical Physics, DOI 10.1007/978-3-319-45784-0_1

Once the Lieb–Robinson bounds for commutators are established, we combine them with results of the theory of strongly continuous semigroups to derive properties of the infinite-volume dynamics. These allow us to extend Lieb–Robinson bounds to time-dependent *multi*-commutators, see Theorems 4.10, 4.11 and 5.4. The new bounds on multi-commutators make possible rigorous studies of dynamical properties that are relevant for response theory of interacting fermion systems. For instance, they yield tree-decay bounds in the sense of [BPH1, Sect. 4] if interactions decay sufficiently fast in space (typically some polynomial decay with large enough degree is needed). In fact, by using the Lieb–Robinson bounds for multi-commutators, we extend in [BP5, BP6] our results [BPH1, BPH2, BPH3, BPH4] on free fermions to interacting particles with short-range interactions. This is an important application of such new bounds: The rigorous microscopic derivation of Ohm and Joule's laws for *interacting* fermions, in the AC-regime. See Chap. 6 and [BP4] for a historical perspective on this subject.

Via Theorems 6.1 and 6.5, we show, for example, how Lieb–Robinson bounds for multi-commutators can be applied to derive decay properties of the so-called *AC-conductivity measure* at high frequencies. This result is new and is obtained in Chap. 6. Cf. [BP5, BP6]. Lieb–Robinson bounds for multi-commutators have, moreover, further applications which go beyond the use on linear response theory presented in Chap. 6. For instance, as explained in Sects. 4.5 and 5.3, they also make possible the study of *non-linear* corrections to linear responses to external perturbations.

The new bounds can also be applied to *non-autonomous systems*. Indeed, the existence of a fundamental solution for the non-autonomous initial value problem related to infinite systems of fermions with time-dependent interactions is usually a non-trivial problem because the corresponding generators are time-dependent unbounded operators. The time-dependency cannot, in general, be isolated into a bounded perturbation around some unbounded time-constant generator and usual perturbation theory cannot be applied. In many important cases, the time-dependent part of the generator is not even relatively bounded with respect to (w.r.t.) the constant part. In fact, no unified theory of non-autonomous evolution equations that gives a complete characterization of the existence of fundamental solutions in terms of properties of generators, analogously to the Hille–Yosida generation theorems for the autonomous case, is available. See, e.g., [K4, C, S, P, BB] and references therein. Note that the existence of a fundamental solution implies the well-posedness of the initial value problem related to states or observables of interacting lattice fermions, provided the corresponding evolution equation has a unique solution for any initial condition.

The Lieb–Robinson bounds on multi-commutators we derive here yield the existence of fundamental solutions as well as other general results on non-autonomous initial value problems related to fermion systems on lattices with interactions which are non-vanishing in the whole space and time-dependent. This is done in a rather constructive way, by considering the large volume limit of finite-volume dynamics, without using standard sufficient conditions for existence of fundamental solutions of non-autonomous linear evolution equations. If interactions decay exponentially fast in space, then we moreover show, also by using Lieb–Robinson bounds on multi-commutators, that the *non-autonomous* dynamics is smooth w.r.t. its generator on the dense set of local observables. See Theorem 5.6. Note that the generator of the (non-autonomous) dynamics generally has, in our case, a time-dependent domain, and the existence of a dense set of smooth vectors is a priori not at all clear.

Observe that the evolution equations for lattice fermions are not of parabolic type, in the precise sense formulated in [AT], because the corresponding generators do not generate analytic semigroups. They seem to be rather related to Kato's hyperbolic case [K2, K3, K4]. Indeed, by structural reasons – more precisely, the fact that the generators are derivations on a C^*-algebra – the time-dependent generator defines a stable family of operators in the sense of Kato. Moreover, this family always possesses a common core. In some specific situations one can directly show that the completion of this core w.r.t. a conveniently chosen norm defines a so-called admissible Banach space $\mathcal{Y}$ of the generator at any time, which satisfies further technical conditions leading to Kato's hyperbolic conditions [K2, K3, K4]. See also [BB, Sect. 5.3.] and [P, Sect.VII.1]. Nevertheless, the existence of such a Banach space $\mathcal{Y}$ is a priori unclear in the general case treated here (Theorem 5.5).

Our central results are Theorems 4.10, 4.11 and 5.4. Other important assertions are Corollary 4.12 and Theorems 5.5, 5.6, 5.8, 5.9, 6.1 and 6.5. The manuscript is organized as follows:

- In order to make our results accessible to a wide audience, in particular to students in Mathematics with little Physics background, Chap. 2 presents basics of Quantum Mechanics, keeping in mind its algebraic formulation.
- Chapter 3 introduces the algebraic setting for fermions, in particular the CAR C^*-algebra. Other standard objects (like fermions, bosons, Fock space, CAR, etc.) of quantum theory are also presented, for pedagogical reasons.
- Chapter 4 is devoted to Lieb–Robinson bounds, which are generalized to multi-commutators. We also give a proof of the existence of the infinite-volume dynamics as well as some applications of such bounds. The tree-decay bounds on time-dependent multi-commutators (Corollary 4.12) are proven here. However, only the autonomous dynamics is considered in this section.
- Chapter 5 extends results of Chap. 4 to the non-autonomous case. We prove, in particular, the existence of a fundamental solution for the non-autonomous initial value problems related to infinite interacting systems of fermions on lattices with time-dependent interactions (Theorem 5.5). This implies well-posedness of the corresponding initial value problems for states and observables, provided their solutions are unique for any initial condition. Applications in (possibly non-linear) response theory (Theorems 5.8, 5.9) are discussed as well.
- Finally, Chap. 6 explains how Lieb–Robinson bounds for multi-commutators can be applied to study (quantum) charged transport properties within the AC-regime. This analysis yields, in particular, the asymptotics at high frequencies of the so-called AC-conductivity measure. See Theorems 6.1 and 6.5.

Notation 1.1
(i) *We denote by D any positive and finite generic constant. These constants do not need to be the same from one statement to another.*

(ii) *A norm on the generic vector space $\mathcal{X}$ is denoted by $\| \cdot \|_\mathcal{X}$ and the identity map of $\mathcal{X}$ by $\mathbf{1}_\mathcal{X}$. The C*-algebra of all bounded linear operators on $(\mathcal{X}, \| \cdot \|_\mathcal{X})$ is denoted by $\mathcal{B}(\mathcal{X})$. The scalar product on a Hilbert space $\mathcal{X}$ is denoted by $\langle \cdot, \cdot \rangle_\mathcal{X}$.*

(iii) *If O is an operator, $\| \cdot \|_O$ stands for the graph norm on its domain.*

(iv) *By a slight abuse of notation, we denote in the sequel elements $X_i \in Y$ depending on the index $i \in I$ by expressions of the form $\{X_i\}_{i \in I} \subset Y$ (instead of $(X_i)_{i \in I} \subset I \times Y$).*

Chapter 2
Algebraic Quantum Mechanics

2.1 Emergence of Quantum Mechanics

The main principles of physics were considered as well-founded by the end of the nineteenth century, even with, for instance, no satisfactory explanation of the phenomenon of thermal radiation, first discovered in 1860 by G. Kirchhoff. In contrast to classical physics, which deals with continuous quantities, Planck's intuition was to introduce an intrinsic discontinuity of energy and a unusual[1] statistics (without any conceptual foundation, in a ad hoc way) to explain thermal radiation in 1900. Assuming the existence of a quantum of action h, the celebrated Planck's constant, and this pivotal statistics he derived the well-known Planck's law of thermal radiation. Inspired by Planck's ideas, Einstein presented his famous discrete (corpuscular) theory of light to explain the photoelectric effect.

Emission spectra of chemical elements had also been known since the nineteenth century and no theoretical explanation was available at that time. It became clear that electrons play a key role in this phenomenon. However, the classical solar system model of the atom failed to explain the emitted or absorbed radiation. Following again Planck's ideas, N. Bohr proposed in 1913 an atomic model based on discrete energies that characterize electron orbits. It became clear that the main principles of classical physics are unable to describe atomic physics.

Planck's quantum of action, Einstein's quanta of light (photons), and Bohr's atomic model could not be a simple extension of classical physics, which, in turn, could also not be questioned in its field of validity. N. Bohr tried during almost a decade to conciliate the paradoxical–looking microscopic phenomena by defining a radically different kind of logic. Bohr's concept of complementarity gave in 1928 a conceptual solution to that problem and revolutionized the usual vision of

---

[1] In regards to Boltzmann's studies, which meanwhile have strongly influenced Planck's work. In modern terms M.K.E.L. Planck used the celebrated Bose–Einstein statistics.

© The Author(s) 2017

J.-B. Bru and W. de Siqueira Pedra, *Lieb-Robinson Bounds for Multi-commutators and Applications to Response Theory*, SpringerBriefs in Mathematical Physics, DOI 10.1007/978-3-319-45784-0_2

nature. See, e.g., [B]. For more details on the emergence of quantum mechanics, see also [R]. Classical logic should be replaced by quantum logic as claimed [BvN] by G. Birkhoff and J. von Neumann in 1936. See also [F].

On the level of theoretical physics, until 1925, quantum corrections were systematically included, in a rather ad hoc manner, into classical theories to allow explicit discontinuous properties. Then, two apparently complementary directions were taken by W.K. Heisenberg and E. Shrödinger, respectively, to establish basic principles of the new quantum physics, in contrast with the "old quantum theory" starting in 1900. Indeed, even with the so-called correspondence principle of N. Bohr, "many problems, even quite central ones like the spectrum of helium atom, proved inaccessible to any solution, no matter how elaborate the conversion", see [R, p. 18].

These parallel theories elaborated almost at the same time were in competition to be the new quantum theory until their equivalence became clear, thanks to J. von Neumann who strongly contributed to the mathematical foundations of Quantum Mechanics in the years following 1926. They are nowadays known in any textbook on Quantum Mechanics as the Schrödinger and Heisenberg pictures of Quantum Mechanics. Schrödinger's view point is generally the most known and refers to the approach we first explain.

2.2 Schrödinger Picture of Quantum Mechanics (S1)

Following de Broglie's studies on (Rutherford–) Bohr's model and Einstein's theory of gases, E. Schrödinger took into account the wave theory of matter in 1925. Indeed, by learning from wave optics in Classical Physics as well as from de Broglie's hypothesis on the wave property of matter, he derived the celebrated *Schrödinger equation*, which describes the time evolution of the wave behavior of all quantum objects. In mathematical words, this time-dependent behavior is described by some family $\{\psi(t)\}_{t\in\mathbb{R}}$ of wave functions within some Hilbert space $\mathcal{H}$, which depends on the quantum system under consideration. This evolution is fixed by a (possibly unbounded) self-adjoint operator $H = H^*$ acting on $\mathcal{H}$: Indeed, for any initial wave function $\psi(0) \in \mathcal{H}$ at $t = 0$, the wave function at arbitrary time $t \in \mathbb{R}$ is uniquely determined by the Schrödinger equation

$$i\partial_t \psi(t) = H\psi(t) , \qquad t \in \mathbb{R} . \tag{2.1}$$

This implies in particular that the time evolution is *unitary*:

$$\psi(t) = e^{-it H}\psi(0) , \qquad t \in \mathbb{R} . \tag{2.2}$$

A typical example is given by $\mathcal{H} = L^2(\mathbb{R}^3)$ with $\psi(0)$ being taken to be a normalized vector of $\mathcal{H}$. In this case, $|\psi(t, x)|^2$ is interpreted as the probability density to detect the quantum particle at time $t \in \mathbb{R}$ and space position $x \in \mathbb{R}^3$.

2.3   Heisenberg Picture of Quantum Mechanics (H2)

Quantities like position, momentum, energy, etc., are represented by *self-adjoint* operators acting on $\mathcal{H}$ and are called *observables*. They refer to all properties of the physical system that can be measured. An important one is of course the energy observable, also named *Hamiltonian*, in reference to the celebrated Hamiltonian mechanics. It is, by definition, the self-adjoint operator H in the Schrödinger equation (2.1).

In this context, the outcomes of measurements of the physical quantity associated with an arbitrary observable B have a random character, the statistical distribution of which is completely described by the family $\{\psi(t)\}_{t \in \mathbb{R}}$ of wave functions solving (2.1). At time $t \in \mathbb{R}$, its expectation value is given by the real number

$$\langle \psi(t), B\psi(t) \rangle_{\mathcal{H}} = \langle \psi(0), e^{itH} B e^{-itH} \psi(0) \rangle_{\mathcal{H}} . \qquad (2.3)$$

See (2.2). Here, $\langle \cdot, \cdot \rangle_{\mathcal{H}}$ denotes the scalar product of $\mathcal{H}$. Viewing the state as time-dependent and the observable fixed, like in Schrödinger's picture of Quantum Mechanics, is equivalent to viewing the state as being fixed and the observable evolving as follows:

$$B \mapsto \tau_t(B) \doteq e^{itH} B e^{-itH} , \qquad t \in \mathbb{R} . \qquad (2.4)$$

The latter refers to Heisenberg's view point: For every bounded Hamiltonians $H \in \mathcal{B}(\mathcal{H})$, the map (2.4) defines a one-parameter continuous group $\{\tau_t\}_{t \in \mathbb{R}}$ acting on $\mathcal{B}(\mathcal{H})$, the Banach space $\mathcal{B}(\mathcal{H})$ of all bounded linear operators on $\mathcal{H}$, and satisfying the (autonomous) evolution equation

$$\forall t \in \mathbb{R} : \quad \partial_t \tau_t = \tau_t \circ \delta = \delta \circ \tau_t , \qquad \tau_0 = \mathbf{1}_{\mathcal{B}(\mathcal{H})} , \qquad (2.5)$$

where δ is the generator defined by

$$\delta(B) \doteq i [H, B] \doteq HB - BH , \qquad B \in \mathcal{B}(\mathcal{H}) . \qquad (2.6)$$

Note that $\{\tau_t\}_{t \in \mathbb{R}}$ is a family of isomorphims of the Banach space $\mathcal{B}(\mathcal{H})$ and, for all $B_1, B_2 \in \mathcal{B}(\mathcal{H})$,

$$\delta(B_1^*) = \delta(B_1)^* \quad \text{and} \quad \delta(B_1 B_2) = \delta(B_1)B_2 + B_1\delta(B_2) . \qquad (2.7)$$

A linear operator δ acting on any algebra with involution (like $\mathcal{B}(\mathcal{H})$, see Sect. 2.5) that satisfies such properties is called *symmetric derivation*. (The symmetry property refers to the first equality.) Indeed, generators of groups of automorphisms of C^*-algebras (Sect. 2.5) are necessarily symmetric derivations.

In this approach the wave function is then fixed for all times. This view point took its origin in Heisenberg's study of the dispersion relation done in 1925. Schrödinger's *wave* mechanics dovetailed with Heisenberg's *matrix* mechanics.

Remark 2.1 (Unbounded Hamiltonians)

If $H = H^*$ is unbounded then it is not clear whether (2.4) defines a C_0-group (that is, a strongly continuous group) $\{\tau_t\}_{t\in\mathbb{R}}$ of automorphisms of $\mathcal{B}(\mathcal{H})$ or not. This fact is, however, not important here. Indeed, one starts (S1) either with Schrödinger's equation and (2.4) is well-defined, (H1) or with a C_0-group $\{\tau_t\}_{t\in\mathbb{R}}$ of automorphisms generated by a (possibly unbounded) symmetric derivation δ, see (2.5) and (2.7). The latter uses the semigroup theory [BR1, EN] and refers to the algebraic formulation of Quantum Mechanics explained in Sect. 2.5.

2.4 Non-autonomous Quantum Dynamics

If $H_t = H_t^*$ is now a time-dependent self-adjoint operator acting on some Hilbert space $\mathcal{H}$ for any time $t \in \mathbb{R}$, the Schrödinger equation

$$i\partial_t\psi(t) = H_t\psi(t) \ , \qquad t \in \mathbb{R} \ ,$$

formally leads to a solution

$$\psi(t) = U_{t,0}\psi(0) \ , \qquad t \in \mathbb{R} \ , \tag{2.8}$$

with $\{U_{t,s}\}_{s,t\in\mathbb{R}}$ being, a priori, the two-parameter group of unitary operators on $\mathcal{H}$ generated by the (anti-self-adjoint) operator $-iH_t$:

$$\forall s, t \in \mathbb{R}: \quad \partial_t U_{t,s} = -iH_t U_{t,s} \ , \quad U_{s,s} \doteq \mathbb{1}_{\mathcal{H}} \ . \tag{2.9}$$

This two-parameter family satisfies the cocycle (Chapman–Kolmogorov) property

$$\forall s, r, t \in \mathbb{R}: \quad U_{t,s} = U_{t,r} U_{r,s} \ . \tag{2.10}$$

Equation (2.9) is a *non-autonomous* evolution equation. The well-posedness of such non-autonomous initial value problems requires some regularity properties of the family $\{H_t\}_{t\in\mathbb{R}}$ of self-adjoint operators. For instance, if $\{H_t\}_{t\in\mathbb{R}} \in C(\mathbb{R}; \mathcal{B}(\mathcal{H}))$ is a continuous family of bounded operators, the existence, uniqueness and even an explicit form of the solution of (2.9) on the space $\mathcal{B}(\mathcal{H})$ (that is, in the norm/uniform topology) is given by the Dyson–Phillips series:

$$U_{t,s} \doteq \mathbb{1}_{\mathcal{H}} + \sum_{k\in\mathbb{N}} (-i)^k \int_s^t ds_1 \cdots \int_s^{s_{k-1}} ds_k \ H_{s_1} \cdots H_{s_k} \ , \qquad s, t \in \mathbb{R} \ . \tag{2.11}$$

In this case, $\{U_{t,s}\}_{s,t\in\mathbb{R}}$ is a norm-continuous two-parameter group of *unitary* operators. In particular, the norm $\|\psi(t)\|_{\mathcal{H}}$ of (2.8) is constant for all times $t \in \mathbb{R}$ and the statistical interpretation of this wave function is still meaningful. Moreover, since the map $B \mapsto B^*$ from $\mathcal{B}(\mathcal{H})$ to $\mathcal{B}(\mathcal{H})$ is continuous (in the norm/uniform topology, see [RS1, Theorem VI.3 (e)]),

$$\forall s, t \in \mathbb{R}: \quad \partial_t \mathrm{U}^*_{t,s} = i \mathrm{U}^*_{t,s} \mathrm{H}_t , \quad \mathrm{U}^*_{s,s} \doteq \mathbf{1}_{\mathcal{H}} . \qquad (2.12)$$

(This property is not that clear in the strong topology since the map $B \mapsto B^*$ is not continuous anymore, but it could still be proven. See as an example [BB, Lemma 68].)

However, the well-posedness of non-autonomous evolution equations like (2.9) is much more delicate for *unbounded* generators. It has been studied, after the first result of Kato in 1953 [K1], for decades by many authors (Kato again [K2, K3] but also Yosida, Tanabe, Kisynski, Hackman, Kobayasi, Ishii, Goldstein, Acquistapace, Terreni, Nickel, Schnaubelt, Caps, Tanaka, Zagrebnov, Neidhardt, etc.), see, e.g., [BB, K4, C, S, P, NZ] and the corresponding references cited therein. Yet, no unified theory of such linear evolution equations that gives a complete characterization analogously to the Hille–Yosida generation theorems [EN] is known.

Assuming the well-posedness of the non-autonomous evolution Eq. (2.9), the expectation value of any observable B (i.e., a self-adjoint operator acting on $\mathcal{H}$) is given, similarly to (2.3), by the real number

$$\langle \psi(t) , B\psi(t) \rangle_{\mathcal{H}} = \langle \psi(s) , \mathrm{U}^*_{t,s} B \mathrm{U}_{t,s} \psi(s) \rangle_{\mathcal{H}} , \qquad s, t \in \mathbb{R} . \qquad (2.13)$$

By (2.9) and (2.12), in the Heisenberg picture of Quantum Mechanics (H2), we observe for any family $\{\mathrm{H}_t\}_{t \in \mathbb{R}} \in C(\mathbb{R}; \mathcal{B}(\mathcal{H}))$ that

$$B \mapsto \tau_{t,s}(B) \doteq \mathrm{U}^*_{t,s} B \mathrm{U}_{t,s} , \qquad s, t \in \mathbb{R} ,$$

defines a two-parameter family $\{\tau_{t,s}\}_{s,t \in \mathbb{R}}$ of automorphisms of $\mathcal{B}(\mathcal{H})$ satisfying the (reverse) cocycle property

$$\forall s, r, t \in \mathbb{R}: \quad \tau_{t,s} = \tau_{r,s} \tau_{t,r} , \qquad (2.14)$$

(cf. (2.10)) as well as the evolution equation

$$\forall s, t \in \mathbb{R}: \quad \partial_t \tau_{t,s} = \tau_{t,s} \circ \delta_t , \quad \tau_{s,s} = \mathbf{1}_{\mathcal{B}(\mathcal{H})} . \qquad (2.15)$$

Here, δ_t is the time-dependent generator defined by

$$\delta_t(B) \doteq i [\mathrm{H}_t, B] \doteq \mathrm{H}_t B - B \mathrm{H}_t , \qquad B \in \mathcal{B}(\mathcal{H}) . \qquad (2.16)$$

Compare with Eqs. (2.5) and (2.6). Equation (2.15) is *another type* of non-autonomous evolution equation on the Banach space $\mathcal{B}(\mathcal{H})$, the well-posedness of which *is much more easier* to prove than the one of (2.9) for unbounded generators.

Indeed, non-autonomous evolution equations in mathematics usually refer to non-autonomous initial value problems

$$\forall t \geq s: \qquad \partial_t \mathfrak{U}_{t,s} = -\mathfrak{G}_t \mathfrak{U}_{t,s} , \quad \mathfrak{U}_{s,s} \doteq \mathbf{1}_{\mathcal{X}} , \qquad (2.17)$$

with generators $\mathfrak{G}_t$ acting on some Banach space $\mathcal{X}$ for times $t \geq s$. One important mathematical issue of (2.9) or (2.17) for unbounded generators is to find sufficient conditions to ensure that $H_t U_{t,s}$ or $\mathfrak{G}_t \mathfrak{U}_{t,s}$ are always well-defined on some (possibly time-dependent) dense subset $\mathfrak{D}$ of $\mathcal{H}$ or $\mathcal{X}$.

This problem does not appear in the non-autonomous initial value problem (2.15). In particular, if the non-autonomous evolution equation

$$\forall s, t \in \mathbb{R}: \qquad \partial_s \tau_{t,s} = -\delta_s \circ \tau_{t,s} , \qquad \tau_{t,t} = \mathbf{1}_{\mathcal{B}(\mathcal{H})} ,$$

is well-posed for some possibly unbounded family $\{\delta_t\}_{t\in\mathbb{R}}$ of generators, then (2.15) is also well-posed, see for instance [BB, Lemma 93]. The converse does *not* hold true, in general. Indeed, in contrast with (2.9) and (2.17), there is no domain conservation in (2.15) to take care even if $\{\delta_t\}_{t\in\mathbb{R}}$ is a family of unbounded generators. An example is given in Sect. 5.1, compare in particular Corollary 5.2 (iii) with Theorem 5.5.

As a consequence, for non-autonomous dynamics the Heisenberg picture of Quantum Mechanics is mathematically more natural or technically advantageous as compared to the Schrödinger picture. This gives a first argument to start the quantum formalism with the Heisenberg picture, instead of the Schrödinger one as it is done in many elementary textbooks on quantum physics. This approach refers to the so-called algebraic formulation of Quantum Mechanics widely used in Quantum Statistical Mechanics and Quantum Field Theory.

2.5 Algebraic Formulation of Quantum Mechanics (H1–S2)

Algebraic Quantum Mechanics is an approach, starting in the forties (cf. GNS construction), which *reverses* the view point presented in Sects. 2.2–2.4 by taking the Heisenberg picture of Quantum Mechanics (H1) as the more fundamental one. Therefore, instead of starting with Hilbert spaces and the Schrödinger equation, one uses C^*-dynamical systems, that is, a pair constituted of a C^*-algebra and a group of $*$-automorphisms. The first generalizes the Banach space $\mathcal{B}(\mathcal{H})$ of all bounded linear operators acting on some Hilbert space $\mathcal{H}$ and the second, the map (2.4). They are defined as follows:

(i): Let $\mathcal{X} \equiv (\mathcal{X}, +, \cdot_{\mathbb{C}})$ be a complex vector space with a product map defined on the Cartesian product $\mathcal{X} \times \mathcal{X}$ by

$$(B_1, B_2) \mapsto B_1 B_2 .$$

$\mathcal{X}$ is an associative and distributive algebra, when, for any $B_1, B_2, B_3 \in \mathcal{X}$ and all complex numbers $\alpha_1, \alpha_2 \in \mathbb{C}$,

$$(B_1 + B_2)B_3 = B_1 B_3 + B_2 B_3 , \quad (B_1 B_2)B_3 = B_1(B_2 B_3) ,$$
$$B_3(B_1 + B_2) = B_3 B_1 + B_3 B_2 , \quad \alpha_1 \alpha_2 (B_1 B_2) = (\alpha_1 B_1)(\alpha_2 B_2) .$$

In the sequel, an algebra carries, by definition, an associative and distributive product. $\mathcal{X}$ is a commutative algebra if $B_1 B_2 = B_2 B_1$ for any $B_1, B_2 \in \mathcal{X}$. $1 \in \mathcal{X}$ is the unit (or identity) of $\mathcal{X}$ when $B\mathbf{1} = \mathbf{1}B = B$ for all $B \in \mathcal{X}$. If $\mathbf{1} \in \mathcal{X}$ exists then it is unique and $\mathcal{X}$ is named a unital algebra.

(ii): An involution is a map $B \mapsto B^*$ from an algebra $\mathcal{X}$ to $\mathcal{X}$ that, by definition, satisfies, for any $B_1, B_2 \in \mathcal{X}$ and $\alpha_1, \alpha_2 \in \mathbb{C}$,

$$(B_1^*)^* = B_1 , \quad (B_1 B_2)^* = B_2^* B_1^* , \quad (\alpha_1 B_1 + \alpha_2 B_2)^* = \overline{\alpha_1} B_1^* + \overline{\alpha_2} B_2^* .$$

An algebra $\mathcal{X}$ equipped with an involution is a $*$-algebra and $B \in \mathcal{X}$ is self-adjoint when $B = B^*$. In this case, by uniqueness of the unit, one checks that a unit $\mathbf{1}$ has to be self-adjoint.

(iii): Let $\| \cdot \|_{\mathcal{X}}$ be a norm on a vector space $\mathcal{X}$. Then, $\mathcal{X} \equiv (\mathcal{X}, \| \cdot \|_{\mathcal{X}})$ is a normed algebra whenever $\mathcal{X}$ is an algebra and

$$\| B_1 B_2 \|_{\mathcal{X}} \leq \| B_1 \|_{\mathcal{X}} \| B_2 \|_{\mathcal{X}} , \qquad B_1, B_2 \in \mathcal{X} .$$

A normed algebra $\mathcal{X}$ is a Banach algebra if $\mathcal{X}$ is complete with respect to (w.r.t.) the norm $\| \cdot \|_{\mathcal{X}}$. A Banach algebra $\mathcal{X}$ equipped with an involution such that

$$\| B \|_{\mathcal{X}} = \| B^* \|_{\mathcal{X}} , \qquad B \in \mathcal{X} ,$$

is a Banach $*$-algebra. Then, a Banach $*$-algebra $\mathcal{X}$ is a C^*-algebra whenever

$$\| B^* B \|_{\mathcal{X}} = \| B \|_{\mathcal{X}}^2 , \qquad B \in \mathcal{X} . \tag{2.18}$$

If $\mathcal{X}$ is a Banach $*$-algebra, then there is a unique norm $\| \cdot \|_{\mathcal{X}}$ on $\mathcal{X}$ such that $(\mathcal{X}, \| \cdot \|_{\mathcal{X}})$ is a C^*-algebra. Note also that in C^*-algebras there is a natural notion of spectrum, which is a real subset for any self-adjoint element.

(iv): Let $\mathcal{X}$ and $\mathcal{Y}$ be two C^*-algebras. A linear map $\pi : \mathcal{X} \to \mathcal{Y}$ is a $*$-homomorphism when it preserves the product and involution of the C^*-algebras, i.e., if, for all $B_1, B_2 \in \mathcal{X}$,

$$\pi (B_1 B_2) = \pi (B_1) \pi (B_2) \quad \text{and} \quad \pi \left(B_1^* \right) = \pi (B_1)^* .$$

Such maps π are automatically contractive [BR1, Proposition 2.3.1] and even isometric when π is injective [BR1, Proposition 2.3.3]. Bijective $*$-homomorphisms are called $*$-isomorphisms. C^*-algebras $\mathcal{X}$ and $\mathcal{Y}$ are said to be $*$-isomorphic whenever there exists a $*$-isomorphism $\pi : \mathcal{X} \to \mathcal{Y}$. $*$-isomorphisms from $\mathcal{X}$ to $\mathcal{X}$ are named $*$-automorphisms of the C^*-algebra $\mathcal{X}$.

For more details on the theory of C^*-algebras, see, e.g., [BR1, KR1, KR2].

A well-known example of unital C^*-algebra is given by the Banach space $\mathcal{B}(\mathcal{H})$ of all bounded linear operators acting on some Hilbert space $\mathcal{H}$. The norm on $\mathcal{B}(\mathcal{H})$ is of course the operator norm, as before, and the involution is defined by taking the adjoint of operators. The complex vector space of complex-valued, measurable, bounded functions on some set equipped with the sup-norm and the point-wise product can also be seen as a unital commutative C^*-algebra.

We are now in position to explain the algebraic approach of Quantum Mechanics, which starts as follows.

Heisenberg Picture of Quantum Mechanics (H1). A physical system is described by its physical properties, i.e., by a non-empty set $\mathcal{O} \neq \emptyset$ of all physical quantifies that can be measured in this system, as well as by the relations between them. Elements $B \in \mathcal{O}$ are called *observables* and are taken as self-adjoint elements of a unital[2] C^*-algebra $\mathcal{X}$. Each self-adjoint element B represents some apparatus (or measuring device) and its spectrum corresponds to all values that can come up by measuring the corresponding physical quantity. The quantum dynamics is given by a C_0-group (that is, a strongly continuous group) $\tau \doteq \{\tau_t\}_{t \in \mathbb{R}}$ of $*$-automorphisms generated [EN, 1.2 Definition, 1.4 Theorem] by a symmetric derivation δ acting on the C^*-algebra $\mathcal{X}$. In particular, by [EN, 1.3 Lemma (ii)], it satisfies the (autonomous) evolution equation

$$\forall t \in \mathbb{R}: \quad \partial_t \tau_t = \tau_t \circ \delta = \delta \circ \tau_t, \qquad \tau_0 = \mathbf{1}_{\mathcal{X}},$$

with δ being a possibly unbounded operator acting on $\mathcal{X}$. Compare with Eqs. (2.5)–(2.7). Recall also that symmetric derivations refer to (linear) operators satisfying properties (2.7) on $\mathcal{X}$. The pair $(\mathcal{X}, \tau)$ is known as a (autonomous) C^*-*dynamical system*. A similar automorphism family can be defined for non-autonomous dynamics by using (2.14) and (2.15) on the domain $\mathrm{Dom}(\delta_t) \subseteq \mathcal{X}$ of a time-dependent symmetric derivation δ_t for $t \in \mathbb{R}$. See for instance Corollary 5.2 (iii). In this case, one speaks about non-autonomous C^*-dynamical systems.

Schrödinger Picture of Quantum Mechanics (S2). States are not anymore defined from a wave function within some Hilbert space, like in Sect. 2.2. States on the C^*-algebra $\mathcal{X}$ are, by definition, continuous linear functionals $\rho \in \mathcal{X}^*$ which are normalized and positive, i.e., $\rho(\mathbf{1}) = 1$ and $\rho(B^*B) \geq 0$ for all $B \in \mathcal{X}$. They represent the state of the physical system. Observe for instance that Eq. (2.3), or (2.13) in the non-autonomous situation, defines a continuous linear functional on the C^*-algebra $\mathcal{B}(\mathcal{H})$ that is positive and normalized, provided $\|\psi(0)\|_{\mathcal{H}} = 1$. Thus, a state ρ represents the statistical distribution of all measures of any observable $B \in \mathcal{X}$. For commutative C^*-algebras, it corresponds to a probability distribution. If $\{\tau_t\}_{t \in \mathbb{R}}$ is a C_0-group of $*$-automorphism of $\mathcal{X}$, then, for any time $t \in \mathbb{R}$ and state $\rho \in \mathcal{X}^*$,

$$\rho_t \doteq \rho \circ \tau_t \in \mathcal{X}^*$$

---

[2]The existence of a unit $\mathbf{1} \in \mathcal{X}$ is assumed to simplify discussions.

is also a state. The same holds true if the dynamics would have been non-autonomous. In the Schrödinger picture, the dynamics is consequently given by the family $\{\rho_t\}_{t\in\mathbb{R}}$ of states.

Therefore, in the algebraic formulation of Quantum Mechanics (H1–S2), there is no a priori Hilbert space structure appearing in the mathematical framework, in contrast with the approach S1–H2 presented in Sects. 2.2–2.4. In fact, by S1–H2 one fixes a unique Hilbert space right from the beginning, whereas the use of H1–S2 can lead to a (not necessarily unique) Hilbert space that depends on the choice of the state.

By [H2, p. 274], I.E. Segal was the first who proposed to leave the Hilbert space approach to consider quantum observables as elements of certain involutive Banach algebras, now known as C^*-algebras. The relation between the algebraic formulation and the usual Hilbert space based formulation of Quantum Mechanics has been established via one important result obtained in the forties: The celebrated *GNS* (Gel'fand–Naimark–Segal) representation of states.

Indeed, by [BR1, Lemma 2.3.10], a positive linear functional ρ over a $*$-algebra $\mathcal{X}$ satisfies

$$\rho(B_1^* B_2) = \overline{\rho(B_2^* B_1)}\,, \qquad B_1,\, B_2 \in \mathcal{X}\,,$$

and the Cauchy–Schwarz inequality:

$$|\rho(B_1^* B_2)|^2 \le \rho(B_1^* B_1)\rho(B_2^* B_2)\,, \qquad B_1,\, B_2 \in \mathcal{X}\,.$$

Therefore, if $\mathcal{X}$ is a unital C^*-algebra and $\rho \in \mathcal{X}^*$ is a state then

$$\mathcal{L}_\rho \doteq \{B \in \mathcal{X} \,:\, \rho(B^* B) = 0\} \tag{2.19}$$

is a closed left-ideal of $\mathcal{X}$, i.e., $\mathcal{L}_\rho$ is a closed subspace such that $\mathcal{X}\mathcal{L}_\rho \subset \mathcal{L}_\rho$, and one can define a scalar product on the quotient $\mathcal{X}/\mathcal{L}_\rho$, which can be completed to get a Hilbert space $\mathcal{H}_\rho$. For more details on the GNS construction, see [BR1, KR1].

The GNS representation has led to very important applications of the Tomita–Takesaki theory (see, e.g., [BR1, KR2]), developed in seventies, to Quantum Field Theory and Statistical Mechanics. These developments mark the beginning of the algebraic approach to Quantum Mechanics and Quantum Field Theory. For more details, see, e.g., [E]. In fact, the algebraic formulation turned out to be extremely important and fruitful for the mathematical foundations of Quantum Statistical Mechanics. See for instance discussions of Sect. 3.5, in particular Lemmata 3.3 and 3.4. In particular, it has been an important branch of research during decades with lots of works on quantum spin and Fermi systems. See, e.g., [BR2, I] (spin) and [AM, BP2, BP3] (Fermi).

2.6   Representation Theory – The Importance of the Algebraic Approach for Infinite Systems

We discuss here how C^*-algebras can be represented by spaces of bounded operators acting Hilbert spaces. A *representation* on the Hilbert space $\mathcal{H}$ of a C^*-algebra $\mathcal{X}$ is, by definition [BR1, Definition 2.3.2], a *-homomorphism π from $\mathcal{X}$ to the unital C^*-algebra $\mathcal{B}(\mathcal{H})$ of all bounded linear operators acting on $\mathcal{H}$. In this case, $\mathcal{H}$ is named the representation (Hilbert) space and if it is finite (resp. infinite) dimensional then we have a finite (resp. infinite) dimensional representation of $\mathcal{X}$. Injective representations are called faithful.

By the Gelfand–Naimark theorem [Dix], each C^*-algebra has, at least, one faithful representation. In particular, since faithful representations are isometric [BR1, Proposition 2.3.3], any C^*-algebra can be identified with some C^*-subalgebra of the C^*-algebra $\mathcal{B}(\mathcal{H})$ of all bounded linear operators acting on some Hilbert space $\mathcal{H}$. In fact, as mentioned in Sect. 2.5, the algebraic formulation of Quantum Mechanics (H1–S2) leads to a Hilbert space $\mathcal{H}_\rho$ for any state ρ, via its GNS representation. A faithful representation can be derived in this way if there exists a state ρ for which $\mathcal{L}_\rho = \{0\}$ (2.19), i.e., if a faithful state exists for the algebra under consideration.

Uniqueness of representations of C^*-algebras is clearly wrong. Indeed, for any representation $\pi : \mathcal{X} \to \mathcal{B}(\mathcal{H})$, we can construct another one by doubling the Hilbert space $\mathcal{H}$ and the map π, via a direct sum $\mathcal{H}_1 \oplus \mathcal{H}_2$ with $\mathcal{H}_1, \mathcal{H}_2$ being two copies of $\mathcal{H}$. Therefore, one uses a notion of "minimal" representations of C^*-algebras: If $\pi : \mathcal{X} \to \mathcal{B}(\mathcal{H})$ is a representation of a C^*-algebra $\mathcal{X}$ on the Hilbert space $\mathcal{H}$, we say that it is *irreducible*, whenever $\{0\}$ and $\mathcal{H}$ are the only closed subspaces of $\mathcal{H}$ which are invariant w.r.t. to any operator of $\pi(\mathcal{X}) \subset \mathcal{B}(\mathcal{H})$.

Now, it is well-known [Na] that if a C^*-algebra $\mathcal{X}$ is isomorphic to the C^*-algebra $\mathcal{K}(\mathcal{H}) \subset \mathcal{B}(\mathcal{H})$ of all compact operators on some Hilbert space $\mathcal{H}$, then, up to unitary equivalence, $\mathcal{X}$ has *only one* irreducible representation (the canonical one on $\mathcal{H}$). The converse is true for *separable* Hilbert spaces: If $\mathcal{X}$ is a C^*-algebra with a faithful representation on a separable Hilbert space $\mathcal{H}$ and if all irreducible representations of $\mathcal{X}$ are unitarily equivalent, then $\mathcal{X}$ is isomorphic to the C^*-algebra $\mathcal{K}(\mathcal{H})$ of compact operators on some Hilbert space $\mathcal{H}$. This result is known as the Rosenberg theorem [Ros]. In other words, one gets the following theorem:

Theorem 2.2 (Uniqueness of irreducible representations – I)
If a C^*-algebra $\mathcal{X}$ has a faithful representation on a separable Hilbert space, then its irreducible representation is unique (up to unitary equivalence) iff $\mathcal{X}$ is isomorphic to some C^*-algebra of compact operators on some Hilbert space.

The question whether all C^*-algebras with a unique (up to unitary equivalence) irreducible representation is isomorphic to an algebra of compact operators on a non-separable Hilbert space is known as "Naimark's problem". Indeed, this question is highly non-trivial. It depends on the continuum hypothesis and not only on the axioms of the Zermelo–Fraenkel set theory with the axiom of choice (ZFC) [Wea, Chap. 19].

In the finite dimensional situation, the C^*-algebra of compact operators is of course equal to the whole C^*-algebra of bounded operators. Therefore, Theorem 2.2 implies the following assertion.

Corollary 2.3 (Uniqueness of irreducible representations – II)
If the C^*-algebra $\mathcal{X}$ is isomorphic to $\mathcal{B}(\mathcal{H})$ for some finite dimensional Hilbert space $\mathcal{H}$, then its irreducible representation is unique, up to unitary equivalence. Any isomorphism $\mathcal{X} \to \mathcal{B}(\mathcal{H})$ of C^*-algebras is such an irreducible representation.

As a consequence, in the finite dimensional situation, the algebraic and Hilbert space based approaches turns out to be equivalent to each other. However, this is not anymore the case in the infinite dimensional situation for unital C^*-algebras because the C^*-algebra of all compact operators *cannot* have a unit:

Corollary 2.4 (Non-uniqueness of irreducible representations)
Any unital C^*-algebra $\mathcal{X}$ with an infinite dimensional faithful representation on a separable Hilbert space has more than one unitarily non-equivalent irreducible representation.

Because of Corollary 2.4, the algebraic approach is more general than the Hilbert space based approach, in the case of infinite dimensional unital underlying C^*-algebras. In condensed matter physics the non-uniqueness of irreducible representations is intimately related to the existence of various thermodynamically stable phases of the same material. Because of this, no reasonable microscopic theory of first order phase transitions is possible within the Hilbert space based approach, and the use of the algebraic setting is imperative.

This fact was first observed by Haag in 1962 [H1], who established that the non-uniqueness of the ground state of the BCS model in infinite volume is related to the existence of several inequivalent irreducible representations [BR1, Definition 2.3.2] of the Hamiltonian, see also [TW, E].

Chapter 3
Algebraic Setting for Interacting Fermions on the Lattice

3.1 Single Quantum Particle on Lattices

All quantum particles carry an intrinsic form of angular momentum, the so–called *spin*, first introduced by W. Pauli in the twenties. It is reflected by a spin quantum number $\mathfrak{s} \in \mathbb{N}/2$ which gives rise to the finite spin set

$$\mathrm{S} \doteq \{-\mathfrak{s}, -\mathfrak{s} + 1, \dots \mathfrak{s} - 1, \mathfrak{s}\} \subset \mathbb{N}. \tag{3.1}$$

In fact, S is the spectrum of the spin observable associated with the quantum particle.

If $\mathfrak{s} \notin \mathbb{N}$ is half–integer then the corresponding particles are named *fermions* while $\mathfrak{s} \in \mathbb{N}$ means by definition that we have *bosons*. For instance, among all elementary particles of the standard model in Particle Physics, quarks and leptons (like electrons, $\mathfrak{s} = 1/2$) are fermions while all the other ones – the gluon, photon, Z– and W– bosons as well as the Higgs bosons – are bosons. Many known composite particles like protons ($\mathfrak{s} = 1/2$) are fermions. Others are bosons, like for instance Helium 4.

By the celebrated spin–statistics theorem, fermionic wave functions are antisymmetric with respect to (w.r.t.) permutations of particles, whereas the bosonic ones are symmetric. In the sequel, we consider the fermionic case which is *only* defined here via the antisymmetry of many–body wave functions (Sect. 3.2), or equivalently by the Canonical Anti–commutation Relations (CAR) in the algebraic formulation (Sect. 3.5). Therefore, in order to simplify notation, we omit the spin property of quantum particles because it is *completely irrelevant* in all our proofs, up to obvious modifications. So, we consider the case $\mathfrak{s} = 0$, i.e., $\mathrm{S} \doteq \{0\}$, even if this is not coherent with the definition explained just above of fermions in Physics.

Additionally, the host material for the quantum particle is a cubic crystal, i.e., a lattice

$$\mathfrak{L} \doteq \mathbb{Z}^d \times \mathrm{S} \equiv \mathbb{Z}^d , \qquad d \in \mathbb{N}.$$

© The Author(s) 2017
J.-B. Bru and W. de Siqueira Pedra, *Lieb-Robinson Bounds for Multi-commutators and Applications to Response Theory*, SpringerBriefs in Mathematical Physics, DOI 10.1007/978-3-319-45784-0_3

This special choice is again not essential in our proofs. In fact, we could take instead of $\mathbb{Z}^d$ any countable metric space $\mathfrak{L}$ as soon as it is regular, as defined in [NS, Sect. 3.1]. See also Sect. 4.2 for more details. (If $\mathfrak{s} \neq 0$ then it would suffice to equip the set $\mathfrak{L} \doteq \mathbb{Z}^d \times S \neq \mathbb{Z}^d$ with the metric of $\mathbb{Z}^d$ while omitting the spin variable.)

Therefore, the one–particle Hilbert space representing the set of all wave functions of any quantum particle on the lattice is given by the space

$$\ell^2(\mathfrak{L}) \doteq \left\{ \psi : \mathfrak{L} \to \mathbb{C} \text{ such that } \sum_{x \in \mathfrak{L}} |\psi(x)|^2 < \infty \right\}$$

of square–summable functions on the lattice $\mathfrak{L}$. Here, the scalar product of $\ell^2(\mathfrak{L})$ is defined by

$$\langle \psi, \varphi \rangle_{\ell^2(\mathfrak{L})} \doteq \sum_{x \in \mathfrak{L}} \overline{\psi(x)} \varphi(x) , \qquad \psi, \varphi \in \ell^2(\mathfrak{L}) .$$

The canonical orthonormal basis of $\ell^2(\mathfrak{L})$ is given by the family $\{\mathfrak{e}_x\}_{x \in \mathfrak{L}}$ defined by

$$\mathfrak{e}_x(y) \doteq \delta_{x,y} , \qquad x, y \in \mathfrak{L}. \tag{3.2}$$

Here, $\delta_{k,l}$ is the Kronecker delta, that is, the function of two variables defined by $\delta_{k,l} = 1$ if $k = l$ and $\delta_{k,l} = 0$ otherwise.

In real systems, the quantum particle is contained in an arbitrary large but finite region. Therefore, we use the notation $\mathcal{P}_f(\mathfrak{L}) \subset 2^{\mathfrak{L}}$ for the set of all *finite* subsets of $\mathfrak{L}$ and we meanwhile denote by

$$\ell^2(\Lambda) \doteq \left\{ \psi \in \ell^2(\mathfrak{L}) : \psi|_{\Lambda^c} = 0 \right\} \subseteq \ell^2(\mathfrak{L}) \tag{3.3}$$

the Hilbert subspace of square–summable functions on any *possibly infinite* sub-set $\Lambda \subseteq \mathfrak{L}$ with complement $\Lambda^c \doteq \mathfrak{L} \backslash \Lambda$. Clearly, the Hilbert subspace $\ell^2(\Lambda)$ has $\{\mathfrak{e}_x\}_{x \in \Lambda}$ as canonical orthonormal basis and, for any $\Lambda \in \mathcal{P}_f(\mathfrak{L})$, its dimension thus equals the volume $|\Lambda|$ of Λ.

Then, as explained in Sect. 2.2, the quantum dynamics is defined by the Schrödinger equation (2.1) for some one–particle Hamiltonian H_1 acting on $\mathcal{H} = \ell^2(\Lambda)$ for any (possibly infinite) subset $\Lambda \subseteq \mathfrak{L}$. A standard example of such self–adjoint operators is given by

$$[H_1(\psi)](x) = \sum_{y \in \mathfrak{L}} h(|x - y|) \psi(y) , \qquad x \in \Lambda, \ \psi \in \ell^2(\Lambda) , \tag{3.4}$$

for any function $h : [0, \infty) \to \mathbb{R}$, the absolute value of which decreases suffi-ciently fast at infinity. This example includes d–dimensional discrete Laplacians, see Sect. 6.2.

3.2 Quantum Many–Body Systems on Lattices

Assume that quantum particles are within some (possibly infinite) subset $\Lambda \subseteq \mathfrak{L}$. A priori, the Hilbert space representing the set of all wave functions of $n \in \mathbb{N}$ identical particles is given by the Hilbert space

$$\ell^2(\Lambda)^{\otimes n} \doteq \ell^2(\Lambda) \otimes \cdots \otimes \ell^2(\Lambda) \, ,$$

the n–fold tensor product of $\ell^2(\Lambda)$ with scalar product defined by

$$\langle \psi_1 \otimes \cdots \otimes \psi_n, \varphi_1 \otimes \cdots \otimes \varphi_n \rangle_{\ell^2(\Lambda)^{\otimes n}} \doteq \langle \psi_1, \varphi_1 \rangle_{\ell^2(\Lambda)} \cdots \langle \psi_n, \varphi_n \rangle_{\ell^2(\Lambda)} \, ,$$

for any $\psi_1, \ldots, \psi_n, \varphi_1, \ldots, \varphi_1 \in \ell^2(\Lambda)$. A canonical orthonormal basis of $\ell^2(\Lambda)^{\otimes n}$ is given by the family

$$\left\{ \mathfrak{e}_{x_1} \otimes \cdots \otimes \mathfrak{e}_{x_n} \right\}_{x_1, \ldots, x_n \in \Lambda} \, , \tag{3.5}$$

where we recall that $\{\mathfrak{e}_x\}_{x \in \Lambda}$ is the (canonical) orthonormal basis of $\ell^2(\Lambda)$ defined by (3.2). Because of (3.3), note that

$$\ell^2(\Lambda)^{\otimes n} \subseteq \ell^2(\mathfrak{L})^{\otimes n}$$

for any $\Lambda \subseteq \mathfrak{L}$ and $n \in \mathbb{N}$.

In Quantum Mechanics, however, quantum particles are *indistinguishable* (or indiscernible), i.e., we cannot distinguish them, even in principle. Indistinguishability is a concept already used in Classical Mechanics, for instance in Botlzmann's 'Combinatorial Approach' to derive the so–called Maxwell–Boltzmann statistics. Two classical objects are indeed indistinguishable when they share the same properties, up to their spatio–temporal location. In particular, by some form of impenetrability assumption, their indistinguishability does not prevent them from being two *different* individuals and so, a spatio–temporal permutation of the two objects yields another physical state.

This property is no longer true in Quantum Mechanics. Quoting E. Schrödinger [Sh]: *"You cannot mark an electron, you cannot paint it red. Indeed, you must not even think of it as marked."* This has an important mathematical consequence on the modelling of composite quantum objects, the individuality of which becomes philosophically questionable. This was implicitly used by M.K.E.L. Planck in his famous study of thermal radiation law, but rather in *ad hoc* way,[1] without conceptual foundations. For more details on that issue, including references, we strongly recommend [FK].

In fact, the expectation value (2.3) of any observable must not depend on the arbitrary numbering of particles. In other words, the wave function $\psi^{(n)} \in \ell^2(\Lambda)^{\otimes n}$ have to satisfy the equality

---

[1] He may have discovered it by working backwards from the thermal radiation law, see [FK, p. 86].

$$\langle \psi^{(n)}, B\psi^{(n)} \rangle_{\ell^2(\Lambda)^{\otimes n}} = \langle S_\pi \psi^{(n)}, BS_\pi \psi^{(n)} \rangle_{\ell^2(\Lambda)^{\otimes n}} \qquad (3.6)$$

for all $B = B^* \in \mathcal{B}(\ell^2(\Lambda)^{\otimes n})$, where $S_\pi \in \mathcal{B}(\ell^2(\mathfrak{L})^{\otimes n})$ is the unitary operator defined for any permutation π of $n \in \mathbb{N}$ elements by the conditions

$$S_\pi (\psi_1 \otimes \cdots \otimes \psi_n) = \psi_{\pi(1)} \otimes \cdots \otimes \psi_{\pi(n)}, \qquad \psi_1, \ldots, \psi_n \in \ell^2(\mathfrak{L}). \qquad (3.7)$$

This yields two drastically different situations:

(b) For any permutation π of $n \in \mathbb{N}$ elements, $S_\pi \psi^{(n)} = \psi^{(n)}$, i.e., $\psi^{(n)}$ is a *symmetric* n–particle wave function. It corresponds to the *boson* case.
(f) For any permutation π of $n \in \mathbb{N}$ elements with sign $(-1)^\pi$, $S_\pi \psi^{(n)} = (-1)^\pi \psi^{(n)}$, i.e., $\psi^{(n)}$ is an *antisymmetric* n–particle wave function. Quantum particles are *fermions*.

Indeed, in contrast with particles with integer spins (boson case), physical particles with half–integer spins (fermion case), obey the Pauli exclusion principle, which says that two identical fermions cannot occupy the same quantum state simultaneously. The latter is reflected in the antisymmetry property of many–fermion wave functions.

Therefore, we mathematically distinguish fermions and bosons only with symmetry properties of wave functions w.r.t. to permutations. In fact, as already mentioned in Sect. 3.1, the spin dependence is, from the technical point of view of proofs, irrelevant here (up to obvious modifications) and without loss of generality (w.l.o.g.) we consider fermions without taking into account its spin in our notation.

Remark 3.1 (Anyons)
By implementing the permutation symmetry property in the configuration space before the "quantization", in the two dimensional space $\mathbb{R}^2$, one has a continuum of (fractional) statistics ranging from the fermionic to the bosonic cases. This refers to the existence of anyons [G, LM, Wi], which has been observed in the context of the fractional quantum Hall effect. Anyons (like bosons as well) do not play any role in the sequel.

Remark 3.2 (Parastatistics)
If $[B, S_\pi] = 0$ then Eq. (3.6) trivially holds true for all states $\psi^{(n)} \in \ell^2(\Lambda)^{\otimes n}$. One could thus assume that, for any permutation π of $n \in \mathbb{N}$ elements, the states $\psi^{(n)}$ and $S_\pi \psi^{(n)}$ cannot be distinguished by any experiment. This view point restricts the set of possible observables to those commuting with all permutation operators S_π. Different statistics, again ranging from the fermionic to the bosonic cases, can then be found from a mathematical perspective. This refers to the so–called parastatistics (which is invariant under the quantum dynamics). Philosophically, this view point has the advantage to restore the individuality of quantum particles, in the classical sense. A historical overview on this approach is given in [FK, Sect. 3.8].

Therefore, for any fixed $n \in \mathbb{N}$, we define the orthogonal projection $P_n \in \mathcal{B}(\ell^2(\mathfrak{L})^{\otimes n})$ onto the subspace of antisymmetric n–particle wave functions in $\ell^2(\mathfrak{L})^{\otimes n}$ by

$$P_n \doteq \frac{1}{n!} \sum_{\pi \in \mathcal{P}_n} (-1)^\pi S_\pi \tag{3.8}$$

with S_π being the operator defined via (3.7) and where

$$\mathcal{P}_n \doteq \{\pi : \{1, \ldots, n\} \to \{1, \ldots, n\} \text{ bijective}\} \tag{3.9}$$

denotes the set of all permutations π of $n \in \mathbb{N}$ elements. Then, for $n \in \mathbb{N}$, the Hilbert space representing the set of all n–fermion wave functions is given by the Hilbert subspace

$$P_n \ell^2(\Lambda)^{\otimes n} \subseteq \ell^2(\Lambda)^{\otimes n}$$

for any (possibly infinite) subset $\Lambda \subseteq \mathfrak{L}$.

As explained in Sect. 2.2, the quantum dynamics is defined by the Schrödinger equation (2.1) for some Hamiltonian $H^{\otimes n}$ acting on $\mathcal{H} = P_n \ell^2(\Lambda)^{\otimes n}$ at fixed $n \in \mathbb{N}$ and $\Lambda \subseteq \mathfrak{L}$. A standard example of such self–adjoint operators is given by

$$[H^{\otimes n}(\psi)](x) = H_1 \otimes \mathbf{1}_{\ell^2(\Lambda)} \cdots \otimes \mathbf{1}_{\ell^2(\Lambda)} + \cdots + \mathbf{1}_{\ell^2(\Lambda)} \otimes \cdots \otimes \mathbf{1}_{\ell^2(\Lambda)} \otimes H_1 + I^{\otimes n} , \tag{3.10}$$

with the one–particle Hamiltonian H_1 defined by (3.4) while $I^{\otimes n}$ is defined by the conditions

$$I^{\otimes n} P_n \left(e_{x_1} \otimes \cdots \otimes e_{x_n} \right) = \sum_{1 \le j < k \le n} v\left(\left| x_j - x_k \right| \right) P_n \left(e_{x_1} \otimes \cdots \otimes e_{x_n} \right)$$

for any $x_1, \ldots, x_n \in \Lambda \subseteq \mathfrak{L}$. $I^{\otimes n}$ represents some interparticle forces which are characterized by a function $v : [0, \infty) \to \mathbb{R}$, the absolute value of which decreases sufficiently fast at infinity.

3.3   Fermion Fock Spaces

In Quantum Statistical Mechanics we are interested in understanding the physical behavior of macroscopic systems from the laws of Quantum Mechanics. This means here that one studies physical properties in the limit $n \to \infty$ of infinite particles. However, the quantum dynamics and even the mathematical framework, that is, the Hilbert space $P_n \ell^2(\Lambda)^{\otimes n}$ of antisymmetric n–particle wave functions, strongly depend on the particle number n, which may additionally be unknown. Moreover, one is often interested in time–dependent particle numbers, as in Quantum Field Theory.

To this end, in 1932 V.A. Fock introduced a space now known as the Fock space defined for Fermi systems by (a priori infinite) direct sums:

$$\mathcal{F}_\Lambda \doteq \bigoplus_{n \in \mathbb{N}_0} P_n \ell^2(\Lambda)^{\otimes n} , \qquad \text{with} \qquad \ell^2(\Lambda)^{\otimes 0} \doteq \mathbb{C} \quad \text{and} \quad P_0 \doteq \mathbf{1}_\mathbb{C} ,$$

for any (possibly infinite) subset Λ of $\mathfrak{L}$. It is an Hilbert space with scalar product defined on $\mathcal{F}_\Lambda \times \mathcal{F}_\Lambda$ by

$$\langle \psi, \varphi \rangle_{\mathcal{F}_\Lambda} \doteq \sum_{n \in \mathbb{N}_0} \langle \psi^{(n)}, \varphi^{(n)} \rangle_{\ell^2(\mathfrak{L})^{\otimes n}} , \qquad \varphi = (\varphi^{(n)})_{n \in \mathbb{N}_0}, \psi = (\psi^{(n)})_{n \in \mathbb{N}_0} \in \mathcal{F}_\Lambda .$$

The element $(1, 0, 0, \ldots) \in \mathcal{F}_\Lambda$ is the zero–particle state, i.e., the so–called *vacuum* vector of the Fock space.

For any *finite* subset $\Lambda \in \mathcal{P}_f(\mathfrak{L})$, recall that $\ell^2(\Lambda)$ (cf. (3.3)) has dimension equal to the volume $|\Lambda|$ of Λ. Therefore, because of the antisymmetry of the n–particle wave function in $\mathcal{F}_\Lambda$,

$$\mathcal{F}_\Lambda = \bigoplus_{n=0}^{|\Lambda|} P_n \ell^2(\Lambda)^{\otimes n} , \qquad \Lambda \in \mathcal{P}_f(\mathfrak{L}). \tag{3.11}$$

Using some elementary combinatorics, one checks in this case that the fermion Fock space $\mathcal{F}_\Lambda$ is a finite dimensional Hilbert space with dimension equal to $2^{|\Lambda|}$ for any $\Lambda \in \mathcal{P}_f(\mathfrak{L})$.

For any possibly infinite subset $\Lambda \subseteq \mathfrak{L}$, the particle number becomes a self–adjoint (possibly unbounded) operator $\mathbf{N}_\Lambda$ defined by

$$(\mathbf{N}_\Lambda \psi)^{(n)} \doteq n \psi^{(n)} , \qquad n \in \mathbb{N}_0 , \tag{3.12}$$

on the domain

$$\text{Dom}(\mathbf{N}_\Lambda) \doteq \left\{ \psi = (\psi^{(n)})_{n \in \mathbb{N}_0} \in \mathcal{F}_\Lambda : \sum_{n \in \mathbb{N}_0} n^2 \langle \psi^{(n)}, \psi^{(n)} \rangle_{\ell^2(\mathfrak{L})^{\otimes n}} < \infty \right\} . \tag{3.13}$$

Any family $\{H^{\otimes n}\}_{n \in \mathbb{N}_0}$ of Hamiltonians acting on $P_n \ell^2(\Lambda)^{\otimes n}$, like those defined by (3.10) for $n \in \mathbb{N}$, gives rise to an operator $\mathbf{H}_\Lambda$ defined for any $n \in \mathbb{N}_0$ by

$$\mathbf{H}_\Lambda \psi^{(n)} \doteq H^{\otimes n} \psi^{(n)} , \qquad \psi^{(n)} \in P_n \ell^2(\Lambda)^{\otimes n} \subset \mathcal{F}_\Lambda . \tag{3.14}$$

It is clearly a symmetric operator on the subspace of $\mathcal{F}_\Lambda$ constituted of sequences that eventually vanish. If $\Lambda \in \mathcal{P}_f(\mathfrak{L})$, it means that $\mathbf{H}_\Lambda$ is self–adjoint, by finite dimensionality of $\mathcal{F}_\Lambda$.

If $\Lambda \subseteq \mathfrak{L}$ is an infinite subset of $\mathfrak{L}$, then $\mathbf{H}_\Lambda$ is closable because it is in any case symmetric, see [RS1, Theorem VIII.1, Sect. 8.2]. By self–adjointness of $H^{\otimes n}$, there is additionally a dense set of analytic vectors [RS1, Sect. X.6] on the subspace of $\mathcal{F}_\Lambda$ constituted of sequences that eventually vanish. Therefore, by Nelson's analytic vector theorem [RS1, Theorem X.39], $\mathbf{H}_\Lambda$ has a self–adjoint closure, again denoted by $\mathbf{H}_\Lambda$.

Therefore, in any case, we obtain from (3.14) a Hamiltonian $\mathbf{H}_\Lambda$ that again defines a quantum dynamics on the Hilbert space $\mathcal{H} = \mathcal{F}_\Lambda$ via the Schrödinger equation (2.1). Typically, such kind of Hamiltonian conserves the particle number in the sense that

$$e^{it\mathbf{H}_\Lambda}\mathbf{N}_\Lambda e^{-it\mathbf{H}_\Lambda} = \mathbf{N}_\Lambda , \qquad t \in \mathbb{R}.$$

In this case, the expectation value of the particle number observable $\mathbf{N}_\Lambda$ w.r.t. any solution of the Schrödinger equation equals

$$\langle \psi (0) , e^{it\mathbf{H}_\Lambda}\mathbf{N}_\Lambda e^{-it\mathbf{H}_\Lambda} \psi (0)\rangle_{\mathcal{F}_\Lambda} = \langle \psi (0) , \mathbf{N}_\Lambda \psi (0)\rangle_{\mathcal{F}_\Lambda} , \qquad \psi (0) \in \mathcal{F}_\Lambda.$$

See Eqs. (2.1), (2.2) and (2.3).

Compared to the first approach described in Sect. 3.2 it is still not clear why the use of the Fock space can be advantageous when the particle number is conserved. The utility of Fock spaces comes from the use of so–called *creation/annihilation operators* explained below.

For more details on Fock spaces, see for instance [BR2, Sect. 5.2.1].

3.4   Creation/Annihilation Operators

Apart from the fact that the Hilbert spaces of Sect. 3.2 depend on the particle number n, one has always to care about combinatorial issues because of the antisymmetry of wave functions. This property of the wave function is encoded in the Fock space in algebraic properties of the so–called creation and annihilation operators $a_x^*, a_x \in \mathcal{B}(\mathcal{F}_\mathfrak{L})$ of a fermion at lattice site $x \in \mathfrak{L}$:

(i): The *annihilation* operator $a_x \in \mathcal{B}(\mathcal{F}_\mathfrak{L})$ of a fermion at lattice site $x \in \mathfrak{L}$ is the (linear) operator uniquely defined by the conditions $a_x(1, 0, 0, \ldots) = 0$ and

$$a_x (P_n (\psi_1 \otimes \cdots \otimes \psi_n)) \doteq \frac{\sqrt{n}}{n!} \sum_{\pi \in \mathcal{P}_n}(-1)^\pi \langle \mathfrak{e}_x, \psi_{\pi(1)}\rangle_{\ell^2(\mathfrak{L})} \psi_{\pi(2)} \otimes \cdots \otimes \psi_{\pi(n)}$$

(3.15)

for any $n \in \mathbb{N}$ and $\psi_1, \ldots, \psi_n \in \ell^2 (\mathfrak{L})$, where we recall that P_n is the orthogonal projection (3.8) onto the subspace of antisymmetric n–particle wave functions and $\mathcal{P}_n$ is the set of all permutations π of n elements, see (3.9).

(ii): The *creation* operator $a_x^* \in \mathcal{B}(\mathcal{F}_\mathfrak{L})$ of a fermion at lattice site $x \in \mathfrak{L}$, which turns out to be the adjoint of a_x, is defined by

$$\left(a_x^*\psi\right)^{(0)} \doteq 0 \qquad \text{and} \qquad \left(a_x^*\psi\right)^{(n)} \doteq \sqrt{n} P_n \left(\psi^{(n-1)} \otimes \mathfrak{e}_x\right) \qquad (3.16)$$

for $n \in \mathbb{N}$, with $\psi = (\psi^{(n)})_{n \in \mathbb{N}_0} \in \mathcal{F}_{\mathfrak{L}}$.

Because of the antisymmetry property, note that, for any $x \in \mathfrak{L}$, $a_x^* a_x^* = 0$, which reflects the Pauli exclusion principle. In fact, straightforward computations show that the family $\{a_x, a_x^*\}_{x \in \mathfrak{L}} \subset \mathcal{B}(\mathcal{F}_{\mathfrak{L}})$ satisfies the celebrated Canonical Anti–commutation Relations (CAR): For any $x, y \in \mathfrak{L}$,

$$a_x a_y + a_y a_x = 0 , \qquad a_x a_y^* + a_y^* a_x = \delta_{x,y} \mathbf{1}_{\mathcal{F}_{\mathfrak{L}}} . \tag{3.17}$$

They are pivotal relations coming from the antisymmetry property of wave functions in the fermion Fock space. For instance, one deduces from (2.18) and (3.17) that in the C^*–algebra $\mathcal{B}(\mathcal{F}_{\mathfrak{L}})$, for any $x \in \mathfrak{L}$,

$$\|a_x\|_{\mathcal{B}(\mathcal{F}_{\mathfrak{L}})}^4 = \|(a_x^* a_x)^2\|_{\mathcal{B}(\mathcal{F}_{\mathfrak{L}})} = \|a_x^* a_x\|_{\mathcal{B}(\mathcal{F}_{\mathfrak{L}})} = \|a_x\|_{\mathcal{B}(\mathcal{F}_{\mathfrak{L}})}^2$$

and since $a_x \neq 0$, we obtain that

$$\|a_x\|_{\mathcal{B}(\mathcal{F}_{\mathfrak{L}})} = \|a_x^*\|_{\mathcal{B}(\mathcal{F}_{\mathfrak{L}})} = 1 , \qquad x \in \mathfrak{L}. \tag{3.18}$$

For more details on creation/annihilation operators, see [BR2, Sect. 5.2.1].

Meanwhile, the particle number operator (3.12)–(3.13) in the (possibly infinite) subset $\Lambda \subseteq \mathfrak{L}$ can be *formally* written on the subspace of antisymmetric n–particle wave functions ($n \in \mathbb{N}$) as

$$\mathbf{N}_\Lambda \psi^{(n)} = \left(\sum_{x \in \Lambda} \mathbf{n}_x \right) \psi^{(n)}, \qquad \psi^{(n)} \in P_n \ell^2(\Lambda)^{\otimes n} ,$$

with

$$\mathbf{n}_x \doteq a_x^* a_x \in \mathcal{B}(\mathcal{F}_{\mathfrak{L}}) \tag{3.19}$$

being the so–called particle number operator at lattice site $x \in \mathfrak{L}$. Because of the CAR (3.17), note that $\mathbf{n}_x$ is a projection operator.

For any finite subset $\Lambda \in \mathcal{P}_f(\mathfrak{L})$, $\mathbf{N}_\Lambda \in \mathcal{B}(\mathcal{F}_\Lambda)$, by finite dimensionality of the local fermion Fock space $\mathcal{F}_\Lambda$ (see (3.11)) and we can see this particle number operator as

$$\mathbf{N}_\Lambda \equiv \sum_{x \in \Lambda} \mathbf{n}_x \in \mathcal{B}(\mathcal{F}_{\mathfrak{L}}). \tag{3.20}$$

In the same way, the operator $\mathbf{H}_\Lambda$ (3.14) can be written in terms of creation and annihilation operators. For instance, if $\mathbf{H}_\Lambda$ is defined from the Hamiltonian (3.4) for any $n \in \mathbb{N}$, then it can be *formally* written on the subspace of antisymmetric n–particle wave functions ($n \in \mathbb{N}$) as

$$\mathbf{H}_\Lambda \psi^{(n)} = \left(\sum_{x,y\in\Lambda} h\left(|x-y|\right) a_x^* a_y + \sum_{x,y\in\Lambda} v(|x-y|)\mathbf{n}_x\mathbf{n}_y \right) \psi^{(n)}$$

for all $\psi^{(n)} \in P_n \ell^2(\Lambda)^{\otimes n}$ and any (possibly infinite) subset $\Lambda \subseteq \mathfrak{L}$.

If $\Lambda \in \mathcal{P}_f(\mathfrak{L})$ is a finite subset then $\mathbf{H}_\Lambda \in \mathcal{B}(\mathcal{F}_\Lambda) \subset \mathcal{B}(\mathcal{F}_\mathfrak{L})$ can be seen as the operator

$$\mathbf{H}_\Lambda \equiv \sum_{x,y\in\Lambda} h\left(|x-y|\right) a_x^* a_y + \sum_{x,y\in\Lambda} v(|x-y|)\mathbf{n}_x\mathbf{n}_y , \qquad (3.21)$$

again by finite dimensionality of $\mathcal{F}_\Lambda$. This formulation of $\mathbf{H}_\Lambda$ can easily be interpreted: The term $a_x^* a_y$ destroy a fermion at lattice site y to create another one at lattice site x. It thus gives rise to fermion transport properties in the physical system and it is related to the (usual) *kinetic terms*. The second term depends on the occupation number (0 or 1) at lattice sites x and y, and yields the interaction energy. It is a so–called *density–density interaction*.

3.5 The Lattice CAR C^*–Algebra

Sections 3.1–3.4 was preliminary sections presenting many–fermion systems on lattices in the usual context of Quantum Mechanics, as explained in Sects. 2.2–2.4. In the sequel, however, we avoid to speak about Hilbert space structures, Fock spaces, etc., by using the algebraic formulation of Quantum Mechanics as explained in Sect. 2.5. To this end, we have to define a C^*–algebra, named the lattice CAR C^*–algebra, defined as follows:

(i): Recall that a_x^*, a_x are the so–called creation and annihilation operators of a fermion at lattice site $x \in \mathfrak{L}$. In Sect. 3.4 we explicitly define them by using the fermion Fock space. In the algebraic approach, however, we only assume the existence of a unit $\mathbf{1}$ and a family $\{a_x, a_x^*\}_{x\in\mathfrak{L}}$ satisfying the CAR: For any $x, y \in \mathfrak{L}$,

$$a_x a_y + a_y a_x = 0 , \qquad a_x a_y^* + a_y^* a_x = \delta_{x,y}\mathbf{1}. \qquad (3.22)$$

Compare with (3.17), by observing that the unit $\mathbf{1}$ refers to the identity map $\mathbf{1}_{\mathcal{F}_\mathfrak{L}}$ of $\mathcal{F}_\mathfrak{L}$. Such commutation relations are indeed sufficient to characterize fermion systems via the Pauli exclusion principle.

(ii): *Every* physical system of fermions on the lattice is associated with some *finite* region Λ of lattice space. The set of all finite subsets of the lattice $\mathfrak{L}$ is denoted by $\mathcal{P}_f(\mathfrak{L}) \subset 2^\mathfrak{L}$. Observables one can then measure on many–fermion systems within any $\Lambda \in \mathcal{P}_f(\mathfrak{L})$ are finite sums of monomials of $\{a_x^*, a_x\}_{x\in\Lambda}$ and $\mathbf{1}$. See (3.19)–(3.21) for explicit examples in the Fock space representation.

For $\Lambda \in \mathcal{P}_f(\mathfrak{L})$, this yields to the *local* CAR C^*–algebra $\mathcal{U}_\Lambda$ as the set of all finite sums of monomials constructed from $\{a_x^*, a_x\}_{x\in\Lambda}$ and the unit $\mathbf{1}$. See Sect. 2.5 for the definition of C^*–algebras. In particular, the particle number operators $\mathbf{n}_x \doteq a_x^* a_x$, $x \in \Lambda$, and the Hamiltonian of the fermion system are self–adjoint elements of $\mathcal{U}_\Lambda$. See again (3.21) for an example of Hamiltonians in the Fock space representation.

Note that one can define annihilation and creation operators of fermions with wave functions $\psi \in \ell^2(\Lambda) \subset \ell^2(\mathfrak{L})$ for any $\Lambda \in \mathcal{P}_f(\mathfrak{L})$ by

$$a(\psi) \doteq \sum_{x \in \Lambda} \overline{\psi(x)} a_x \in \mathcal{U}_\Lambda \, , \quad a^*(\psi) \doteq \sum_{x \in \Lambda} \psi(x) a_x^* \in \mathcal{U}_\Lambda. \qquad (3.23)$$

Clearly, $a^*(\psi) = a(\psi)^*$ for all $\psi \in \ell^2(\Lambda)$ and on the canonical orthonormal basis $\{\mathfrak{e}_x\}_{x \in \mathfrak{L}}$ (3.2), $a(\mathfrak{e}_x) = a_x$ at all $x \in \Lambda$. The map $\psi \mapsto a(\psi)$ (resp. $\psi \mapsto a^*(\psi)$) from $\ell^2(\Lambda)$ to $\mathcal{U}_\Lambda$ is anti–linear (resp. linear) and because of (3.22),

$$a(\psi)a(\varphi) + a(\varphi)a(\psi) = 0 \, , \quad a(\psi)a(\varphi)^* + a(\varphi)^* a(\psi) = \langle \psi, \varphi \rangle_{\ell^2(\mathfrak{L})} \mathbf{1} \quad (3.24)$$

for any $\varphi, \psi \in \ell^2(\Lambda) \subset \ell^2(\mathfrak{L})$. These CAR are a generalization of (3.22).

The relation with the local fermion Fock space $\mathcal{F}_\Lambda$ (3.11) is now clear:

Lemma 3.3 (CAR algebras and fermion Fock spaces for finite systems) *For any $\Lambda \in \mathcal{P}_f(\mathfrak{L})$, the local CAR C^*–algebra $\mathcal{U}_\Lambda$ is $*$–isomorphic to the C^*–algebra $\mathcal{B}(\mathcal{F}_\Lambda)$ of bounded operators acting on $\mathcal{F}_\Lambda$. In particular, its dimension equals $2^{2|\Lambda|}$.*

Proof On the one hand, since $\ell^2(\Lambda)$ has dimension $|\Lambda|$, for any $\Lambda \in \mathcal{P}_f(\mathfrak{L})$, we infer from (3.23)–(3.24) and [BR2, Theorem 5.2.5] that $\mathcal{U}_\Lambda$ is isomorphic to the C^*–algebra of $2^{|\Lambda|} \times 2^{|\Lambda|}$ complex matrices. $\mathcal{U}_\Lambda$ has in particular dimension equal to $2^{2|\Lambda|}$. On the other hand, as explained below Eq. (3.11), for any $\Lambda \in \mathcal{P}_f(\mathfrak{L})$, the dimension of the Fock space $\mathcal{F}_\Lambda$ is equal to $2^{|\Lambda|}$. Therefore, $\mathcal{U}_\Lambda$ is $*$–isomorphic to $\mathcal{B}(\mathcal{F}_\Lambda)$. ∎

Lemma 3.3 yields a *faithful* representation

$$\pi_\Lambda : \mathcal{U}_\Lambda \to \mathcal{B}(\mathcal{F}_\Lambda) \qquad (3.25)$$

of the local CAR C^*–algebra $\mathcal{U}_\Lambda$ on the representation (Hilbert) space $\mathcal{F}_\Lambda$ for every $\Lambda \in \mathcal{P}_f(\mathfrak{L})$. This representation is said to be *canonical* when it maps a_x^*, a_x to creation and annihilation operators defined by (3.15) and (3.16) on $\mathcal{F}_\Lambda$ for any $x \in \mathfrak{L}$. The (canonical) representation is *irreducible* and named the Fock representation. For more details on the representation theory of C^*–algebras, see Sect. 2.6.

Because of Corollary 2.3, one can equivalently use the Fock space formulation within the usual context of Quantum Mechanics or the algebraic approach, provided $\Lambda \in \mathcal{P}_f(\mathfrak{L})$. Compare indeed Lemma 3.3 with the Heisenberg picture of Quantum Mechanics described in Sect. 2.3. This fact is *not anymore* true in the infinite-volume situation, i.e., for *infinite* subsets $\Lambda \subseteq \mathfrak{L}$, since the corresponding Fock space $\mathcal{F}_\Lambda$ has then infinite dimension. See Corollary 2.4.

(iii): Physical systems become macroscopic when they belong to finite regions Λ of lattice that become arbitrarily large. Therefore, we consider a family of cubic boxes[2] defined, for all $L \in \mathbb{R}_0^+$, by

$$\Lambda_L \doteq \{(x_1, \ldots, x_d) \in \mathfrak{L} : |x_1|, \ldots, |x_d| \leq L\} \in \mathcal{P}_f(\mathfrak{L}) . \tag{3.26}$$

Hence, $\{\mathcal{U}_{\Lambda_L}\}_{L \in \mathbb{R}_0^+}$ is an increasing net of C^*–algebras and the set

$$\mathcal{U}_0 \doteq \bigcup_{L \in \mathbb{R}_0^+} \mathcal{U}_{\Lambda_L} \tag{3.27}$$

of local elements is a normed $*$–algebra with $\|A\|_{\mathcal{U}_0} = \|A\|_{\mathcal{U}_{\Lambda_L}}$ for all $A \in \mathcal{U}_{\Lambda_L}$ and $L \in \mathbb{R}_0^+$.

(iv): For physical macroscopic systems one considers the limit "$\Lambda \to \mathfrak{L}$". This is named the thermodynamic limit and one gets an infinite fermion system. This approach yields the CAR C^*–algebra $\mathcal{U}$ of the infinite system, which is by definition the completion of the normed $*$–algebra $\mathcal{U}_0$. It is separable, by finite dimensionality of $\mathcal{U}_\Lambda$ for any $\Lambda \in \mathcal{P}_f(\mathfrak{L})$. In other words, $\mathcal{U}$ is the inductive limit of the finite dimensional C^*–algebras $\{\mathcal{U}_\Lambda\}_{\Lambda \in \mathcal{P}_f(\mathfrak{L})}$. In this construction, $\mathcal{U}_0 \subset \mathcal{U}$ can be seen as the smallest normed $*$–algebra containing all generators $\{a_x\}_{x \in \mathfrak{L}}$.

By replacing Λ with $\mathfrak{L}$ in Eq. (3.23) one can again define annihilation and creation operators $a(\psi)$, $a^*(\psi)$ of fermions with wave functions $\psi \in \ell^2(\mathfrak{L})$. These operators are still well–defined. Indeed, because of the CAR (3.22),

$$\|a(\psi)\|_{\mathcal{U}}^2 = \|a^*(\psi)\|_{\mathcal{U}}^2 = \langle \psi, \psi \rangle_{\ell^2(\mathfrak{L})} , \qquad \psi \in \ell^2(\mathfrak{L}).$$

Compare with (3.18). Hence, the anti–linear (resp. linear) map $\psi \mapsto a(\psi)$ (resp. $\psi \mapsto a^*(\psi)$) from $\ell^2(\mathfrak{L})$ to $\mathcal{U}$ is norm–continuous. Again, $a^*(\psi) = a(\psi)^*$ for all $\psi \in \ell^2(\mathfrak{L})$ and the CAR (3.24) can be extended to all $\varphi, \psi \in \ell^2(\mathfrak{L})$.

The faithful and irreducible canonical representation π_Λ defined by (3.25) gives rise in the infinite-volume limit to a unique faithful and irreducible (canonical) representation $\pi_\mathfrak{L}$ of the CAR C^*–algebra $\mathcal{U}$ such that

$$\pi_\mathfrak{L}(\mathcal{U}_\Lambda) = \pi_\Lambda(\mathcal{U}_\Lambda) = \mathcal{B}(\mathcal{F}_\Lambda) , \qquad \Lambda \in \mathcal{P}_f(\mathfrak{L}). \tag{3.28}$$

This representation is in fact defined via the unique $*$–isomorphisms $\pi_{\{x\}}$, $x \in \mathfrak{L}$, mapping $a_x \in \mathcal{U}$ to the operator $\pi_{\{x\}}(a_x) \equiv a_x \in \mathcal{B}(\mathcal{F}_\mathfrak{L})$ defined by (3.15). It is again named the Fock representation.

The Fock space $\mathcal{F}_\mathfrak{L}$ has infinite dimension and is separable. Thus, by Corollary 2.4, we emphasize again that the Fock space formulation within the usual context of Quantum Mechanics and the algebraic approach are not equivalent to each other.

---

[2]It is a technically convenient choice to define the thermodynamic limit, but one could also take other Van Hove nets. See for instance [BP2, Remark 1.3].

Again by Corollary 2.4, the C^*–algebra $\mathcal{B}(\mathcal{F}_{\mathfrak{L}})$ has more than one unitarily non–equivalent irreducible representation as well. Moreover, $\mathcal{U}$ is in some sense *strictly smaller* than $\mathcal{B}(\mathcal{F}_{\mathfrak{L}})$:

Lemma 3.4 (CAR algebras and fermion Fock spaces for infinite systems)

$$\pi_{\mathfrak{L}}(\mathcal{U}) = \overline{\bigcup_{L \in \mathbb{R}_0^+} \mathcal{B}(\mathcal{F}_{\Lambda_L})} \subsetneq \mathcal{B}(\mathcal{F}_{\mathfrak{L}}).$$

Proof For any non–vanishing $\theta \in \mathbb{R}/(2\pi\mathbb{Z})$, $e^{i\theta \mathbf{N}_{\mathfrak{L}}} \in \mathcal{B}(\mathcal{F}_{\mathfrak{L}})$ but $\pi_{\mathfrak{L}}^{-1}(e^{i\theta \mathbf{N}_{\mathfrak{L}}}) = \emptyset$, with the particle number operator $\mathbf{N}_{\mathfrak{L}}$ being the self–adjoint operator defined by (3.12)–(3.13) for $\Lambda = \mathfrak{L}$. Indeed, for all $\Lambda \in \mathcal{P}_f(\mathfrak{L})$ with complement $\Lambda^c \doteq \mathfrak{L}\backslash\Lambda$ and any function $\psi \in \ell^2(\mathfrak{L}) \subset \mathcal{F}_{\mathfrak{L}}$, a direct computation shows that

$$\left\| \left(e^{i\theta \mathbf{N}_{\mathfrak{L}}} - e^{i\theta \mathbf{N}_{\Lambda}}\right) \psi \right\|_{\mathcal{F}_{\mathfrak{L}}}^2 = \left| e^{i\theta} - 1 \right|^2 \sum_{x \in \Lambda^c} |\psi(x)|^2. \tag{3.29}$$

Here, $\mathbf{N}_{\Lambda} \in \mathcal{B}(\mathcal{F}_{\mathfrak{L}})$ is the particle number operator defined by (3.20) for $\Lambda \in \mathcal{P}_f(\mathfrak{L})$. Therefore, $e^{i\theta \mathbf{N}_{\Lambda}}$ does not converge in $\mathcal{B}(\mathcal{F}_{\mathfrak{L}})$ (norm topology) to $e^{i\theta \mathbf{N}_{\mathfrak{L}}}$ for $\theta \neq 0$.

Assume now that $\pi_{\mathfrak{L}}^{-1}(e^{i\theta \mathbf{N}_{\mathfrak{L}}}) \neq \emptyset$. Then, by Lemma 3.3 and density of $\mathcal{U}_0$ in $\mathcal{U}$, there are two families $\{\Lambda_n\}_{n \in \mathbb{N}} \subset \mathcal{P}_f(\mathfrak{L})$ and $\{U_{\Lambda_n}\}_{n \in \mathbb{N}}$ such that $U_{\Lambda_n} \in \mathcal{B}(\mathcal{F}_{\Lambda_n}) \subset \mathcal{B}(\mathcal{F}_{\mathfrak{L}})$ converges in $\mathcal{B}(\mathcal{F}_{\mathfrak{L}})$ to $e^{i\theta \mathbf{N}_{\mathfrak{L}}}$, as $n \to \infty$. From this and (3.29), one deduces that $\left(U_{\Lambda_n} - e^{i\theta \mathbf{N}_{\Lambda_n}}\right)$ must converge, as $n \to \infty$, to zero in $\mathcal{B}(\mathcal{F}_{\mathfrak{L}})$. The latter is not possible, otherwise $e^{i\theta \mathbf{N}_{\Lambda}}$ would then converge in $\mathcal{B}(\mathcal{F}_{\mathfrak{L}})$ to $e^{i\theta \mathbf{N}_{\mathfrak{L}}}$.

Therefore, for any non–vanishing $\theta \in \mathbb{R}/(2\pi\mathbb{Z})$, $\pi_{\mathfrak{L}}^{-1}(e^{i\theta \mathbf{N}_{\mathfrak{L}}}) = \emptyset$. The assertion then follows, by Eqs. (3.27) and (3.28). ∎

(v): For any non–vanishing $\theta \in \mathbb{R}/(2\pi\mathbb{Z})$, the unitary operator $e^{i\theta \mathbf{N}_{\mathfrak{L}}} \notin \pi_{\mathfrak{L}}(\mathcal{U})$ (see proof of Lemma 3.4) gives rise to a $*$–automorphism

$$B \mapsto e^{i\theta \mathbf{N}_{\mathfrak{L}}} \, B \, e^{-i\theta \mathbf{N}_{\mathfrak{L}}}$$

of $\mathcal{B}(\mathcal{F}_{\mathfrak{L}})$ defined via (3.12) and (3.13). One says that the unitary operator $e^{i\theta \mathbf{N}_{\mathfrak{L}}} \in \mathcal{B}(\mathcal{F}_{\mathfrak{L}})$ implements a *global gauge transformation*, see for instance [BP1, Eq. (A.4)]. A similar $*$–automorphism exists on the CAR C^*–algebra $\mathcal{U}$: For any $\theta \in \mathbb{R}/(2\pi\mathbb{Z})$, it is the unique $*$–automorphism σ_θ of $\mathcal{U}$ satisfying the conditions

$$\sigma_\theta(a_x) = e^{-i\theta} a_x, \qquad x \in \mathfrak{L}. \tag{3.30}$$

Note indeed that, using the Fock representation, one verifies that

$$\pi_{\mathfrak{L}}(\sigma_\theta(B)) = e^{i\theta \mathbf{N}_{\mathfrak{L}}} \, \pi_{\mathfrak{L}}(B) \, e^{-i\theta \mathbf{N}_{\mathfrak{L}}}, \qquad B \in \mathcal{U},$$

for any $\theta \in \mathbb{R}/(2\pi\mathbb{Z})$.

A special role is played by σ_π: Elements B_1, $B_2 \in \mathcal{U}$ satisfying $\sigma_\pi(B_1) = B_1$ and $\sigma_\pi(B_2) = -B_2$ are respectively called *even* and *odd*, while elements $B \in \mathcal{U}$ satisfying $\sigma_\theta(B) = B$ for any $\theta \in \mathbb{R}/(2\pi\mathbb{Z})$ are called *gauge invariant*. The set

$$\mathcal{U}^+ \doteq \{B \in \mathcal{U} \ : \ B = \sigma_\pi(B)\} \subset \mathcal{U} \tag{3.31}$$

of all even elements and the set

$$\mathcal{U}^\circ \doteq \bigcap_{\theta \in \mathbb{R}/(2\pi\mathbb{Z})} \{B \in \mathcal{U} \ : \ B = \sigma_\theta(B)\} \subset \mathcal{U}^+ \tag{3.32}$$

of all gauge invariant elements are $*$–algebras. By continuity of σ_θ, it follows that $\mathcal{U}^+$ and $\mathcal{U}^\circ$ are closed and hence C^*–algebras. $\mathcal{U}^\circ$ is known as the fermion *observable* algebra because it equals the C^*–algebra of all self–adjoint elements of $\mathcal{U}$.

3.6   Lattice Fermi Versus Quantum Spin Systems

Quantum spin systems are models used to describe quantum phenomena appearing at low temperatures in condensed matter physics. They are nowadays particularly important in Quantum Information Theory. This subject appeared right from the beginning, with the emergence of Quantum Mechanics in the twenties. A concise introduction on its history is given in the paper [N], see also the corresponding references therein.

For completeness, we shortly recall that quantum spin systems are infinite systems composed of elementary finite dimensional spaces, originally referring to a spin variable (see (3.1)). Therefore, they are constructed in a similar way as lattice Fermi systems. Mathematically speaking, they are defined via the algebraic formulation of Quantum Mechanics from the so–called *spin* C^*–algebra $\mathcal{Q}$:

(i): With any lattice site $x \in \mathfrak{L}$ we associate a finite dimensional Hilbert space $\mathcal{H}_x \equiv \mathbb{C}^N$ for some $N \in \mathbb{N}$. Typically, the parameter N is the cardinal $|S|$ of the spin set S (3.1). Then, the algebra of local observables over $\Lambda \in \mathcal{P}_f(\mathfrak{L})$ is the subset of self–adjoint elements of the C^*–algebra

$$\mathcal{Q}_\Lambda \doteq \bigotimes_{x \in \Lambda} \mathcal{B}(\mathcal{H}_x) \equiv \mathcal{B}\left(\bigotimes_{x \in \Lambda} \mathcal{H}_x\right).$$

Recall that $\mathcal{B}(\mathcal{H}_x)$ denotes the C^*–algebra of bounded linear operators on $\mathcal{H}_x$ for $x \in \mathfrak{L}$. The dimension of $\mathcal{Q}_\Lambda$ is equal to $N^{2|\Lambda|}$ for any $\Lambda \in \mathcal{P}_f(\mathfrak{L})$. Compare with the local CAR C^*–algebra $\mathcal{U}_\Lambda$, see in particular Lemma 3.3.

(ii): For all $\Lambda^{(1)}$, $\Lambda^{(2)} \in \mathcal{P}_f(\mathfrak{L})$ with $\Lambda^{(1)} \subset \Lambda^{(2)}$, there is a (canonical) isometric inclusion $\mathcal{Q}_{\Lambda^{(1)}} \hookrightarrow \mathcal{Q}_{\Lambda^{(2)}}$ defined by

$$A \mapsto A \otimes \bigotimes_{x \in \Lambda^{(2)} \setminus \Lambda^{(1)}} 1_{\mathcal{H}_x} \, .$$

In particular, using the sequence of cubic boxes defined by (3.26) we observe that $\{\mathcal{Q}_{\Lambda_L}\}_{L \in \mathbb{R}_0^+}$ is also an increasing net of C^*–algebras. Compare with the family $\{\mathcal{U}_{\Lambda_L}\}_{L \in \mathbb{R}_0^+}$.

(iii): Hence, the set

$$\mathcal{Q}_0 \doteq \bigcup_{L \in \mathbb{R}_0^+} \mathcal{Q}_{\Lambda_L}$$

of local elements is a normed $*$–algebra with $\|A\|_{\mathcal{Q}_0} = \|A\|_{\mathcal{Q}_{\Lambda_L}}$ for all $A \in \mathcal{Q}_{\Lambda_L}$ and $L \in \mathbb{R}_0^+$. Compare with the normed $*$–algebra $\mathcal{U}_0$ of local elements defined by (3.27).

For any finite subsets $\Lambda^{(1)}, \Lambda^{(2)} \in \mathcal{P}_f(\mathfrak{L})$ with $\Lambda^{(1)} \cap \Lambda^{(2)} = \emptyset$ we observe that

$$[B_1, B_2] \doteq B_1 B_2 - B_2 B_1 = 0 \, , \qquad B_1 \in \mathcal{Q}_{\Lambda^{(1)}}, \ B_2 \in \mathcal{Q}_{\Lambda^{(2)}}.$$

Because of the CAR (3.22), such a property is also satisfied for all *even* local elements $B_1 \in \mathcal{U}_{\Lambda^{(1)}} \cap \mathcal{U}^+$ and $B_2 \in \mathcal{U}_{\Lambda^{(2)}} \cap \mathcal{U}^+$, see (3.31). However, it is wrong in general for Fermi systems. For instance, the CAR (3.22) trivially yield $[a_x, a_y] = 2a_x a_y$ for any $x, y \in \mathfrak{L}$.

(iv): The *spin* C^*–algebra $\mathcal{Q}$ of the lattice $\mathfrak{L}$ is by definition the completion of the normed $*$–algebra $\mathcal{Q}_0$. It is separable, by finite dimensionality of $\mathcal{Q}_\Lambda$ for $\Lambda \in \mathcal{P}_f(\mathfrak{L})$. In other words, $\mathcal{Q}$ is the inductive limit of the finite dimensional C^*–algebras $\{\mathcal{Q}_\Lambda\}_{\Lambda \in \mathcal{P}_f(\mathfrak{L})}$. Compare with the CAR C^*–algebra $\mathcal{U}$.

Infinite-volume dynamics is then constructed via Lieb–Robinson bounds, as done in Chaps. 4 and 5 for CAR C^*–algebras. Here, we focus on lattice Fermi systems which are, from a technical point of view, slightly more difficult because of the non–commutativity of their elements on different lattice sites, as explained above. However, all the results presented in Chaps. 4 and 5 hold true for quantum spin systems, by restricting them on the C^*–algebra $\mathcal{U}^+ \subset \mathcal{U}$ (3.31) of all even elements and then by replacing $\mathcal{U}^+$ with the spin C^*–algebras $\mathcal{Q}$.

Chapter 4
Lieb–Robinson Bounds for Multi–commutators

Lieb–Robinson bounds for multi–commutators are studied here for fermion systems, only. In the case of quantum spin systems, $\mathcal{U}$ has to be replaced by the infinite tensor product $\mathcal{Q}$ of copies of some finite dimensional C^*–algebra attached to each site $x \in \mathfrak{L}$. See Sect. 3.6. All results of this section also hold in this situation. We concentrate our attention on fermion algebras in view of applications to microscopic foundations of the theory of electrical conduction [BP4, BP5]. Moreover, as explained in Sect. 3.6, the fermionic case is, technically speaking, more involved, because of the non–commutativity of elements of the CAR algebra $\mathcal{U}$ sitting on different lattice sites.

4.1 Interactions and Finite-Volume Dynamics

Following the algebraic formulation of Quantum Mechanics (Sect. 2.5), we have to define a C_0–group (that is, a strongly continuous group) $\{\tau_t\}_{t \in \mathbb{R}}$ of $*$–automorphisms of the CAR C^*–algebra $\mathcal{U}$. On the other hand, as explained in Sect. 3.5, every physical system of particles belongs to some finite region Λ_L (3.26) of lattice space, and they become macroscopic when $L \to \infty$. Therefore, we define the C_0–group $\{\tau_t\}_{t \in \mathbb{R}}$ as a limit $L \to \infty$ of finite-volume dynamics.

We thus need to define a family of Hamiltonians $H_L \in \mathcal{U}_{\Lambda_L}$ for $L \in \mathbb{R}_0^+$. This is done by using the notions of *interactions* and *potentials* defined as follows:

- Interactions are by definition families $\Psi = \{\Psi_\Lambda\}_{\Lambda \in \mathcal{P}_f(\mathfrak{L})}$ of even (cf. (3.31)) and self–adjoint local elements $\Psi_\Lambda = \Psi_\Lambda^* \in \mathcal{U}^+ \cap \mathcal{U}_\Lambda$ with $\Psi_\emptyset = 0$. Obviously, the set of all interactions can be endowed with a real vector space structure:

$$(\alpha_1 \Phi + \alpha_2 \Psi)_\Lambda \doteq \alpha_1 \Phi_\Lambda + \alpha_2 \Psi_\Lambda$$

© The Author(s) 2017
J.-B. Bru and W. de Siqueira Pedra, *Lieb-Robinson Bounds for Multi-commutators and Applications to Response Theory*, SpringerBriefs in Mathematical Physics, DOI 10.1007/978-3-319-45784-0_4

for any interactions Φ, Ψ, and any real numbers α_1, $\alpha_2 \in \mathbb{R}$.

- By potential, we mean here a collection $\mathbf{V} \doteq \{\mathbf{V}_{\{x\}}\}_{x \in \mathfrak{L}}$ of even (cf. (3.31)) and self–adjoint elements such that $\mathbf{V}_{\{x\}} = \mathbf{V}_{\{x\}}^* \in \mathcal{U}^+ \cap \mathcal{U}_{\{x\}}$ for all $x \in \mathfrak{L}$. Such objects are sometimes called *on–site interactions*. Indeed, strictly speaking, a potential is nothing but a special case of interaction. But, the use of this special notion allows us to treat latter the cases for which (4.10) holds true.

Take now any interaction Ψ and potential $\mathbf{V}$. With such objects we associate the (internal) energy observable or Hamiltonian

$$H_L \doteq \sum_{\Lambda \subseteq \Lambda_L} \Psi_\Lambda + \sum_{x \in \Lambda_L} \mathbf{V}_{\{x\}} , \qquad L \in \mathbb{R}_0^+ , \tag{4.1}$$

of the cubic box Λ_L defined by (3.26).

Then, similar to Eqs. (2.4)–(2.6), the finite–volume dynamics corresponds to the continuous group $\{\tau_t^{(L)}\}_{t \in \mathbb{R}}$ of $*$–automorphisms of $\mathcal{U}$ defined by

$$\tau_t^{(L)}(B) = e^{itH_L} B e^{-itH_L} , \qquad B \in \mathcal{U} , \tag{4.2}$$

for any $L \in \mathbb{R}_0^+$, interaction Ψ and potential $\mathbf{V}$. Obviously, its generator is the bounded linear operator $\delta^{(L)}$ defined on $\mathcal{U}$ by

$$\delta^{(L)}(B) \doteq i \sum_{\Lambda \subseteq \Lambda_L} [\Psi_\Lambda, B] + i \sum_{x \in \Lambda_L} [\mathbf{V}_{\{x\}}, B] , \qquad B \in \mathcal{U} . \tag{4.3}$$

It is a symmetric derivation on $\mathcal{U}$ because, for all $B_1, B_2 \in \mathcal{U}$,

$$\delta^{(L)}(B_1^*) = \delta^{(L)}(B_1)^* \quad \text{and} \quad \delta^{(L)}(B_1 B_2) = \delta^{(L)}(B_1) B_2 + B_1 \delta^{(L)}(B_2) .$$

Compare with Eq. (2.7).

Using two functions $h, v : [0, \infty) \to \mathbb{R}$, note that the finite-volume Hamiltonian (4.1) associated with the interaction $\Psi^{(h,v)}$ defined by

$$\Psi_\Lambda^{(h,v)} \doteq h\left(|x - y|\right) a_x^* a_y + \left(1 - \delta_{x,y}\right) h\left(|x - y|\right) a_y^* a_x \tag{4.4}$$
$$+ v\left(|x - y|\right) \left(a_y^* a_y a_x^* a_x + \left(1 - \delta_{x,y}\right) a_x^* a_x a_y^* a_y\right)$$

whenever $\Lambda = \{x, y\}$ for $x, y \in \mathfrak{L}$, and $\Psi_\Lambda^{(h,v)} \doteq 0$ otherwise, is equal in this case to

$$H_L = \sum_{x,y \in \Lambda_L} h\left(|x - y|\right) a_x^* a_y + \sum_{x,y \in \Lambda_L} v(|x - y|) \mathbf{n}_x \mathbf{n}_y , \qquad L \in \mathbb{R}_0^+ .$$

Compare with (3.21) for $\Lambda = \Lambda_L$. This gives a very important – albeit very specific – example of a Fermi model on the lattice. For instance, it includes the celebrated Hubbard model widely used in Physics. Other examples are given in Sect. 6.2.

4.2   Banach Spaces of Short–Range Interactions

The finite-volume dynamics we define in Sect. 4.1 should converge to an infinite-volume one to be able to understand macroscopic systems. In other words, the limit $L \to \infty$ of the continuous group $\{\tau_t^{(L)}\}_{t \in \mathbb{R}}$ of $*$–automorphisms defined by (4.2) has to converge to a C_0–group $\{\tau_t\}_{t \in \mathbb{R}}$ of $*$–automorphisms of the CAR C^*–algebra $\mathcal{U}$. In order to ensure that property (cf. Sect. 4.3), we define Banach spaces of short–range interactions by introducing specific norms for interactions, taking into account space decay.

Following [NOS, Eqs. (1.3)–(1.4)], we consider positive–valued and non–increasing decay functions $\mathbf{F} : \mathbb{R}_0^+ \to \mathbb{R}^+$ satisfying the following properties:

- *Summability on* $\mathfrak{L}$.

$$\|\mathbf{F}\|_{1,\mathfrak{L}} \doteq \sup_{y \in \mathfrak{L}} \sum_{x \in \mathfrak{L}} \mathbf{F}(|x - y|) = \sum_{x \in \mathfrak{L}} \mathbf{F}(|x|) < \infty . \qquad (4.5)$$

- *Bounded convolution constant.*

$$\mathbf{D} \doteq \sup_{x,y \in \mathfrak{L}} \sum_{z \in \mathfrak{L}} \frac{\mathbf{F}(|x - z|)\,\mathbf{F}(|z - y|)}{\mathbf{F}(|x - y|)} < \infty . \qquad (4.6)$$

In the case $\mathfrak{L}$ would be a general countable set with infinite cardinality and some metric d, the existence of such a function $\mathbf{F}$ satisfying (4.5)–(4.6) with d$(\cdot, \cdot)$ instead of $|\cdot - \cdot|$ refers to the so–called *regular* property of $\mathfrak{L}$. For any $d \in \mathbb{N}$, $\mathfrak{L} \doteq \mathbb{Z}^d$ is in this sense regular with the metric d$(\cdot, \cdot) = |\cdot - \cdot|$. Indeed, a typical example of such a $\mathbf{F}$ for $\mathfrak{L} = \mathbb{Z}^d$, $d \in \mathbb{N}$, and the metric induced by $|\cdot|$ is the function

$$\mathbf{F}(r) \doteq (1 + r)^{-(d+\epsilon)} , \qquad r \in \mathbb{R}_0^+ , \qquad (4.7)$$

which has convolution constant $\mathbf{D} \leq 2^{d+1+\epsilon} \|\mathbf{F}\|_{1,\mathfrak{L}}$ for $\epsilon \in \mathbb{R}^+$. See [NOS, Eq. (1.6)] or [Si, Example 3.1]. Note that the exponential function $\mathbf{F}(r) = e^{-\varsigma r}$, $\varsigma \in \mathbb{R}^+$, satisfies (4.5) but not (4.6). Nevertheless, for every function $\mathbf{F}$ with bounded convolution constant (4.6) and any strictly positive parameter $\varsigma \in \mathbb{R}^+$, the function

$$\tilde{\mathbf{F}}(r) = e^{-\varsigma r} \mathbf{F}(r) , \qquad r \in \mathbb{R}_0^+ ,$$

clearly satisfies Assumption (4.6) with a convolution constant that is no bigger than the one of $\mathbf{F}$. In fact, as observed in [Si, Sect. 3.1], the multiplication of such a function $\mathbf{F}$ with a non–increasing weight $f : \mathbb{R}_0^+ \to \mathbb{R}^+$ satisfying $f(r + s) \geq f(r)f(s)$ (logarithmically superadditive function) does not increase the convolution constant $\mathbf{D}$. In the sequel, (4.5)–(4.6) are assumed to be satisfied.

The function $\mathbf{F}$ encodes the short–range property of interactions. Indeed, an interaction Ψ is said to be *short–range* if

$$\|\Psi\|_{\mathcal{W}} \doteq \sup_{x,y \in \mathfrak{L}} \sum_{\Lambda \in \mathcal{P}_f(\mathfrak{L}),\ \Lambda \supset \{x,y\}} \frac{\|\Psi_\Lambda\|_{\mathcal{U}}}{\mathbf{F}(|x-y|)} < \infty. \tag{4.8}$$

Since the map $\Psi \mapsto \|\Psi\|_{\mathcal{W}}$ defines a norm on interactions, the space of short–range interactions w.r.t. to the decay function $\mathbf{F}$ is the real separable Banach space $\mathcal{W} \equiv (\mathcal{W}, \|\cdot\|_{\mathcal{W}})$ of all interactions Ψ with $\|\Psi\|_{\mathcal{W}} < \infty$. Note that a short–range interaction $\Psi \in \mathcal{W}$ is not necessarily weak away from the origin of $\mathfrak{L}$: Generally, the element $\Psi_{x+\Lambda}, x \in \mathfrak{L}$, does not vanish when $|x| \to \infty$. It turns out that all short–range interactions $\Psi \in \mathcal{W}$ define, in a natural way, infinite–volume quantum dynamics, i.e., they define C^*–dynamical systems on $\mathcal{U}$. For more details, see Sect. 4.3, in particular Theorem 4.8. (Recall that C^*–dynamical systems are defined in Sect. 2.5.)

Remark 4.1 (Lattice Fermi models)
The interaction $\Psi^{(h,v)}$ defined in Sect. 4.1 (see (4.4)) belongs to $\mathcal{W}$ as soon as $h, v : [0, \infty) \to \mathbb{R}$ are real–valued and summable functions satisfying

$$\sup_{r \in \mathbb{R}_0^+} \left\{ \frac{|h(r)|}{\mathbf{F}(r)} \right\} < \infty \qquad \text{and} \qquad \sup_{r \in \mathbb{R}_0^+} \left\{ \frac{|v(r)|}{\mathbf{F}(r)} \right\} < \infty. \tag{4.9}$$

Remark 4.2 (Quantum spin models)
All important spin models with no mean field term can be constructed from short–range interactions, as defined above. As examples, we can mention the Ising model, the (quantum) Heisenberg model, the XXZ model, the XY model, the XXZ model, the model [AKLT], etc. See for instance [N] and references therein.

4.3   Existence of Dynamics and Lieb–Robinson Bounds

In Sect. 4.2, we define a Banach space $\mathcal{W}$ of short–range interactions by using a convenient norm $\|\cdot\|_{\mathcal{W}}$ for interactions, see (4.8). $\Psi \in \mathcal{W}$ ensures the existence of an infinite–volume derivation δ associated with Ψ by taking the thermodynamic limit $L \to \infty$ of commutators involving Ψ_Λ, $\Lambda \in \mathcal{P}_f(\mathfrak{L})$, see (4.3). This also holds true for all potentials $\mathbf{V} \doteq \{\mathbf{V}_{\{x\}}\}_{x \in \mathfrak{L}}$, as defined in Sect. 4.1. Every generator of a C^*–dynamical system is a derivation, but the converse does not generally hold. We show here that δ is the generator of a C^*–dynamical system in $\mathcal{U}$ when $\Psi \in \mathcal{W}$ and for all potentials $\mathbf{V} \doteq \{\mathbf{V}_{\{x\}}\}_{x \in \mathfrak{L}}$. Note that the interaction representing $\mathbf{V}$ can possibly be outside $\mathcal{W}$ because we allow $\mathbf{V}$ to be unbounded, i.e., the case

$$\sup_{x \in \mathfrak{L}} \|\mathbf{V}_{\{x\}}\|_{\mathcal{U}} = \infty \tag{4.10}$$

is included in the discussion below.

The key ingredient in this analysis are the so–called *Lieb–Robinson bounds*. Indeed, they lead, among other things, to the existence of the infinite–volume dynamics for interacting particles. By using this, we define a C^*–dynamical system in $\mathcal{U}$ for any short–range interaction $\Psi \in \mathcal{W}$. These bounds are, moreover, a pivotal ingredient to study transport properties of interacting fermion systems later on. Thus, for the reader's convenience, below we review this topic in detail.

It is convenient to introduce at this point the notation

$$S_\Lambda(\tilde{\Lambda}) \doteq \left\{ \mathcal{Z} \subset \Lambda : \mathcal{Z} \cap \tilde{\Lambda} \neq 0 \text{ and } \mathcal{Z} \cap \tilde{\Lambda}^c \neq 0 \right\} \tag{4.11}$$

for any set $\tilde{\Lambda} \subset \Lambda \subset \mathfrak{L}$ with complement $\tilde{\Lambda}^c \doteq \mathfrak{L} \backslash \tilde{\Lambda}$, as well as

$$\partial_\Psi \Lambda \doteq \{x \in \Lambda : \exists \mathcal{Z} \in S_\mathfrak{L}(\Lambda) \text{ with } x \in \mathcal{Z} \text{ and } \Psi_\mathcal{Z} \neq 0\}$$

for any interaction $\Psi \doteq \{\Psi_\mathcal{Z}\}_{\mathcal{Z} \in \mathcal{P}_f(\mathfrak{L})}$ and any finite subset $\Lambda \in \mathcal{P}_f(\mathfrak{L})$ of $\mathfrak{L}$. We are now in position to prove Lieb–Robinson bounds for finite–volume fermion systems with short–range interactions and in presence of potentials:

Theorem 4.3 (Lieb–Robinson bounds)
Let $\Psi \in \mathcal{W}$ and $\mathbf{V}$ be any potential. Then, for any $t \in \mathbb{R}$, $L \in \mathbb{R}_0^+$, and elements $B_1 \in \mathcal{U}^+ \cap \mathcal{U}_{\Lambda^{(1)}}$, $B_2 \in \mathcal{U}_{\Lambda^{(2)}}$ with $\Lambda^{(1)}, \Lambda^{(2)} \in \mathcal{P}_f(\mathfrak{L})$ and $\Lambda^{(1)} \cap \Lambda^{(2)} = \emptyset$,

$$\left\| \left[\tau_t^{(L)}(B_1), B_2 \right] \right\|_\mathcal{U} \leq 2\mathbf{D}^{-1} \|B_1\|_\mathcal{U} \|B_2\|_\mathcal{U} \left(e^{2\mathbf{D}|t| \|\Psi\|_\mathcal{W}} - 1 \right) \tag{4.12}$$

$$\times \sum_{x \in \partial_\Psi \Lambda^{(1)}} \sum_{y \in \Lambda^{(2)}} \mathbf{F}(|x - y|).$$

The constant $\mathbf{D} \in \mathbb{R}^+$ is defined by (4.6).

Proof The arguments are essentially the same as those proving [NS, Theorem 2.3.] for quantum spin systems. Here, we consider fermion systems and we give the detailed proof for completeness and to prepare its extension to time–dependent interactions and potentials, in Theorem 5.1 (i). We fix $L \in \mathbb{R}_0^+$, $B_1 \in \mathcal{U}^+ \cap \mathcal{U}_{\Lambda^{(1)}}$ and $B_2 \in \mathcal{U}_{\Lambda^{(2)}}$ with disjoint sets $\Lambda^{(1)}, \Lambda^{(2)} \subsetneq \Lambda_L$. [Note that $\Lambda^{(1)} \cap \Lambda^{(2)} = \emptyset$ yields $L \geq 1$.]

Let

$$C_{B_2}(\Lambda; t) \doteq \sup_{B \in \mathcal{U}^+ \cap \mathcal{U}_\Lambda, B \neq 0} \frac{\left\| \left[\tau_t^{(L)}(B), B_2 \right] \right\|_\mathcal{U}}{\|B\|_\mathcal{U}}, \qquad t \in \mathbb{R}, \ \Lambda \in \mathcal{P}_f(\mathfrak{L}).$$

At time $t = 0$, we observe that

$$\left| C_{B_2}(\Lambda; 0) \right| \leq 2 \|B_2\|_\mathcal{U} \mathbf{1}\left[\Lambda \cap \Lambda^{(2)} \neq \emptyset\right],$$

while, for any $t \in \mathbb{R}$,

$$C_{B_2}(\Lambda; t) = \sup_{B \in \mathcal{U}^+ \cap \mathcal{U}_\Lambda, B \neq 0} \frac{\left\| [\tau_t^{(L)} \circ \tau_{-t}^{(\Lambda)}(B), B_2] \right\|_{\mathcal{U}}}{\|B\|_{\mathcal{U}}}.$$

Here, $\{\tau_t^{(\Lambda)}\}_{t \in \mathbb{R}}$ is the continuous group of $*$–automorphisms defined like $\{\tau_t^{(L)}\}_{t \in \mathbb{R}}$ by replacing the box Λ_L with the (finite) set $\Lambda \in \mathcal{P}_f(\mathfrak{L})$.

Consider the function

$$f(t) \doteq \left[\tau_t^{(L)} \circ \tau_{-t}^{(\Lambda^{(1)})}(B_1), B_2 \right], \qquad t \in \mathbb{R}. \tag{4.13}$$

Then, using $B_1 \in \mathcal{U}^+ \cap \mathcal{U}_{\Lambda^{(1)}}$ and $\Lambda^{(1)} \subset \Lambda_L$, we deduce from (4.3) and explicit computations that

$$\partial_t f(t) = i \sum_{Z \in \mathcal{S}_{\Lambda_L}(\Lambda^{(1)})} \left[\tau_t^{(L)}(\Psi_Z), f(t) \right] \tag{4.14}$$

$$-i \sum_{Z \in \mathcal{S}_{\Lambda_L}(\Lambda^{(1)})} \left[\tau_t^{(L)} \circ \tau_{-t}^{(\Lambda^{(1)})}(B_1), \left[\tau_t^{(L)}(\Psi_Z), B_2 \right] \right].$$

Let $\mathfrak{g}_t(B)$ be the solution of

$$\forall t \geq 0: \qquad \partial_t \mathfrak{g}_t(B) = i \sum_{Z \in \mathcal{S}_{\Lambda_L}(\Lambda^{(1)})} [\tau_t^{(L)}(\Psi_Z), \mathfrak{g}_t(B)], \qquad \mathfrak{g}_0(B) = B \in \mathcal{U}.$$

Since $\|\mathfrak{g}_t(B)\|_{\mathcal{U}} = \|B\|_{\mathcal{U}}$ for any $B \in \mathcal{U}$, it follows from (4.14), by variation of constants, that

$$\|f(t)\|_{\mathcal{U}} \leq \|f(0)\|_{\mathcal{U}} + 2 \|B_1\|_{\mathcal{U}} \sum_{Z \in \mathcal{S}_{\Lambda_L}(\Lambda^{(1)})} \int_0^{|t|} \left\| [\tau_{\pm s}^{(L)}(\Psi_Z), B_2] \right\|_{\mathcal{U}} ds. \tag{4.15}$$

[The sign of s in $\pm s$ depends whether t is positive or negative.] Hence, as $\Lambda^{(1)}$, $\Lambda^{(2)}$ are disjoint, for any $t \in \mathbb{R}$,

$$C_{B_2}(\Lambda^{(1)}; t) \leq 2 \sum_{Z \in \mathcal{S}_{\Lambda_L}(\Lambda^{(1)})} \|\Psi_Z\|_{\mathcal{U}} \int_0^{|t|} C_{B_2}(Z; \pm s) ds. \tag{4.16}$$

By estimating $C_{B_2}(Z; s)$ in a similar manner and iterating this procedure, we show that, for every $L \in \mathbb{R}_0^+$, $t \in \mathbb{R}$ and all $B_1 \in \mathcal{U}^+ \cap \mathcal{U}_{\Lambda^{(1)}}$, $B_2 \in \mathcal{U}_{\Lambda^{(2)}}$ with disjoint $\Lambda^{(1)}, \Lambda^{(2)} \subset \Lambda_L$,

$$C_{B_2}(\Lambda^{(1)}; t) \leq 2 \|B_2\|_{\mathcal{U}} \sum_{k \in \mathbb{N}} \frac{|2t|^k}{k!} u_k, \tag{4.17}$$

where, for any $k \in \mathbb{N}$,

$$u_k \doteq \sum_{\mathcal{Z}_1 \in \mathcal{S}_{\Lambda_L}(\Lambda^{(1)})} \sum_{\mathcal{Z}_2 \in \mathcal{S}_{\Lambda_L}(\mathcal{Z}_1)} \cdots \sum_{\mathcal{Z}_k \in \mathcal{S}_{\Lambda_L}(\mathcal{Z}_{k-1})} \mathbf{1}\left[\mathcal{Z}_k \cap \Lambda^{(2)} \neq \emptyset\right] \prod_{j=1}^{k} \left\| \Psi_{\mathcal{Z}_j} \right\|_{\mathcal{U}} \,.$$

The above series is absolutely and uniformly convergent for $L \in \mathbb{R}_0^+$ (with fixed $\Lambda^{(1)}, \Lambda^{(2)} \subsetneq \Lambda_L$). Indeed, from straightforward estimates,

$$u_k \leq \mathbf{D}^{k-1} \|\Psi\|_{\mathcal{W}}^k \sum_{x \in \partial_\Psi \Lambda^{(1)}} \sum_{y \in \Lambda^{(2)}} \mathbf{F}(|x - y|) \,, \tag{4.18}$$

by Eqs. (4.6) and (4.8).

Note that (4.17)–(4.18) yield (4.12), provided $\Lambda^{(1)}, \Lambda^{(2)} \subsetneq \Lambda_L$. This last condition can easily be removed by taking, at any fixed $L \in \mathbb{R}_0^+$, an interaction $\tilde{\Psi}^{(L)} \in \mathcal{W}$ defined by $\tilde{\Psi}_{\mathcal{Z}}^{(L)} \doteq \Psi_{\mathcal{Z}}$ for any $\mathcal{Z} \subseteq \Lambda_L$, while $\tilde{\Psi}_{\mathcal{Z}}^{(L)} \doteq 0$ when $\mathcal{Z} \nsubseteq \Lambda_L$. Indeed, for all $L \in \mathbb{R}_0^+$, we obviously have $\|\tilde{\Psi}^{(L)}\|_{\mathcal{W}} \leq \|\Psi\|_{\mathcal{W}}$. Furthermore, for all $L, \tilde{L} \in \mathbb{R}_0^+$ with $\tilde{L} > L$, $\tilde{\tau}_t^{(\tilde{L})} = \tau_t^{(L)}$, where $\{\tilde{\tau}_t^{(\tilde{L})}\}_{t \in \mathbb{R}}$ is the (finite–volume) group of $*$–automorphisms of $\mathcal{U}$ defined by (4.2) with $L = \tilde{L}$ and $\Psi = \tilde{\Psi}^{(L)}$. Therefore, it suffices to apply (4.17)–(4.18) to the interaction $\tilde{\Psi}^{(L)}$ for sufficiently large $\tilde{L} \in \mathbb{R}_0^+$ in order to get the assertion without the condition $\Lambda^{(1)}, \Lambda^{(2)} \subsetneq \Lambda_L$. ∎

As explained in [NS, Theorem 3.1] for quantum spin systems, Lieb–Robinson bounds lead to the existence of the infinite–volume dynamics:

Lemma 4.4 (Finite–volume dynamics as a Cauchy sequence)
Let $\Psi \in \mathcal{W}$ and $\mathbf{V}$ be any potential. Then, for any $t \in \mathbb{R}$, $\Lambda \in \mathcal{P}_f(\mathfrak{L})$, $B \in \mathcal{U}_\Lambda$ and $L_1, L_2 \in \mathbb{R}_0^+$ with $\Lambda \subset \Lambda_{L_1} \subsetneq \Lambda_{L_2}$,

$$\left\| \tau_t^{(L_2)}(B) - \tau_t^{(L_1)}(B) \right\|_{\mathcal{U}} \leq 2 \|B\|_{\mathcal{U}} \|\Psi\|_{\mathcal{W}} |t| \, e^{4\mathbf{D}|t| \|\Psi\|_{\mathcal{W}}}$$

$$\times \sum_{y \in \Lambda_{L_2} \setminus \Lambda_{L_1}} \sum_{x \in \Lambda} \mathbf{F}(|x - y|) \,.$$

Proof Again, the arguments are those proving [NS, Theorem 3.1] for quantum spin systems. We give them for completeness, having also in mind the extension of the lemma to time–dependent interactions and potentials, in Theorem 5.1 (ii). We fix in all the proof $\Lambda \in \mathcal{P}_f(\mathfrak{L})$ and $B \in \mathcal{U}_\Lambda$.

For any $L \in \mathbb{R}_0^+$ and $s, t \in \mathbb{R}$, define the unitary element

$$\mathbf{U}_L(t, s) \doteq e^{it\mathbf{V}_{\Lambda_L}} e^{-i(t-s)H_L} e^{-is\mathbf{V}_{\Lambda_L}} \in \mathcal{U}_{\Lambda_L} \tag{4.19}$$

with

$$\mathbf{V}_{\mathcal{Z}} \doteq \sum_{x \in \mathcal{Z}} \mathbf{V}_{\{x\}} \in \mathcal{U}^+ \cap \mathcal{U}_{\mathcal{Z}} \,, \qquad \mathcal{Z} \in \mathcal{P}_f(\mathfrak{L}) \,.$$

Clearly, $\mathbf{U}_L(t, t) = \mathbf{1}_{\mathcal{U}}$ for all $t \in \mathbb{R}$ while

$$\partial_t \mathbf{U}_L(t,s) = -i G_L(t) \mathbf{U}_L(t,s) \quad \text{and} \quad \partial_s \mathbf{U}_L(t,s) = i \mathbf{U}_L(t,s) G_L(s)$$

with

$$G_L(t) \doteq \sum_{\mathcal{Z} \subseteq \Lambda_L} e^{it \mathbf{V}_{\Lambda_L}} \Psi_{\mathcal{Z}} e^{-it \mathbf{V}_{\Lambda_L}} .$$

Let

$$\tilde{\tau}_t^{(L)}(B) \doteq \mathbf{U}_L(0,t) B \mathbf{U}_L(t,0) , \qquad B \in \mathcal{U}_\Lambda .$$

For any $t \in \mathbb{R}$ and $L \in \mathbb{R}_0^+$ such that $\Lambda \subset \Lambda_L$,

$$\tau_t^{(L)}(B) = \tilde{\tau}_t^{(L)}\left(e^{it \mathbf{V}_{\Lambda_L}} B e^{-it \mathbf{V}_{\Lambda_L}}\right) = \tilde{\tau}_t^{(L)}\left(e^{it \mathbf{V}_\Lambda} B e^{-it \mathbf{V}_\Lambda}\right)$$

and it suffices to study the net $\{\tilde{\tau}_t^{(L)}(B)\}_{L \in \mathbb{R}_0^+}$ in $\mathcal{U}$. The equality above is related to the so–called "interaction picture" (w.r.t. potentials) of the time–evolution defined by the $*$–automorphism $\tau_t^{(L)}$.

Fix $L_1, L_2 \in \mathbb{R}_0^+$ with $\Lambda \subset \Lambda_{L_1} \subsetneq \Lambda_{L_2}$. Note that, for any $t \in \mathbb{R}$,

$$\tilde{\tau}_t^{(L_2)}(B) - \tilde{\tau}_t^{(L_1)}(B) = \int_0^t \partial_s \left\{ \mathbf{U}_{L_2}(0,s) \mathbf{U}_{L_1}(s,t) B \mathbf{U}_{L_1}(t,s) \mathbf{U}_{L_2}(s,0) \right\} ds . \tag{4.20}$$

Straightforward computations yield

$$\partial_s \left\{ \mathbf{U}_{L_2}(0,s) \mathbf{U}_{L_1}(s,t) B \mathbf{U}_{L_1}(t,s) \mathbf{U}_{L_2}(s,0) \right\}$$
$$= i \mathbf{U}_{L_2}(0,s) \left[G_{L_2}(s) - G_{L_1}(s) , \mathbf{U}_{L_1}(s,t) B \mathbf{U}_{L_1}(t,s) \right] \mathbf{U}_{L_2}(s,0)$$
$$= i \mathbf{U}_{L_2}(0,s) e^{is \mathbf{V}_{\Lambda_{L_1}}} \left[B_s, \tau_{t-s}^{(L_1)}(\tilde{B}_t) \right] e^{-is \mathbf{V}_{\Lambda_{L_1}}} \mathbf{U}_{L_2}(s,0) , \tag{4.21}$$

where, for any $s, t \in \mathbb{R}$, we define

$$B_s \doteq e^{-is \mathbf{V}_{\Lambda_{L_1}}} \left(G_{L_2}(s) - G_{L_1}(s) \right) e^{is \mathbf{V}_{\Lambda_{L_1}}} \quad \text{and} \quad \tilde{B}_t \doteq e^{-it \mathbf{V}_\Lambda} B e^{it \mathbf{V}_\Lambda} . \tag{4.22}$$

Thus, we infer from Eqs. (4.20)–(4.22) that

$$\left\| \tilde{\tau}_t^{(L_2)}(B) - \tilde{\tau}_t^{(L_1)}(B) \right\|_{\mathcal{U}} \leq \int_0^{|t|} \left\| \left[\tau_{\pm s - t}^{(L_1)}(B_{\pm s}) , \tilde{B}_t \right] \right\|_{\mathcal{U}} ds . \tag{4.23}$$

[The sign of s in $\pm s$ depends whether t is positive or negative.] Note that $\tilde{B}_t \in \mathcal{U}_\Lambda$ and

$$B_s = \sum_{\mathcal{Z} \subseteq \Lambda_{L_2}, \ \mathcal{Z} \cap (\Lambda_{L_2} \setminus \Lambda_{L_1}) \neq \emptyset} e^{is \mathbf{V}_{\Lambda_{L_2} \setminus \Lambda_{L_1}}} \Psi_{\mathcal{Z}} e^{-is \mathbf{V}_{\Lambda_{L_2} \setminus \Lambda_{L_1}}} \in \mathcal{U}^+ \cap \mathcal{U}_{\Lambda_{L_2}}$$

where, for any $\mathcal{Z} \subseteq \Lambda_{L_2}$,

$$e^{isV_{\Lambda_{L_2}\setminus\Lambda_{L_1}}}\Psi_Z e^{-isV_{\Lambda_{L_2}\setminus\Lambda_{L_1}}}\in\mathcal{U}_Z\,.$$

Now, we apply the Lieb–Robinson bounds given by Theorem 4.3 to deduce that, for any $\Lambda\in\mathcal{P}_f(\mathfrak{L})$, $s,t\in\mathbb{R}$, $B\in\mathcal{U}_\Lambda$ and $L_1,L_2\in\mathbb{R}_0^+$ with $\Lambda\subset\Lambda_{L_1}\subsetneq\Lambda_{L_2}$,

$$\frac{\left\|\left[\tau_{s-t}^{(L_1)}(B_s),\tilde{B}_t\right]\right\|_{\mathcal{U}}}{2\,\|B\|_{\mathcal{U}}}\leq\mathbf{D}^{-1}\left(e^{2\mathbf{D}|s-t|\,\|\Psi\|_{\mathcal{W}}}-1\right) \tag{4.24}$$

$$\times\sum_{\substack{Z\subseteq\Lambda_{L_2},\\ Z\cap(\Lambda_{L_2}\setminus\Lambda_{L_1})\neq\emptyset,\ Z\cap\Lambda=\emptyset}}\|\Psi_Z\|_{\mathcal{U}}\sum_{z\in\partial_\Psi Z}\sum_{x\in\Lambda}\mathbf{F}(|x-z|)$$

$$+\sum_{\substack{Z\subseteq\Lambda_{L_2},\\ Z\cap(\Lambda_{L_2}\setminus\Lambda_{L_1})\neq\emptyset,\ Z\cap\Lambda\neq\emptyset}}\|\Psi_Z\|_{\mathcal{U}}\,.$$

Direct estimates using (4.6) and (4.8) show that

$$\sum_{\substack{Z\subseteq\Lambda_{L_2},\ Z\cap(\Lambda_{L_2}\setminus\Lambda_{L_1})\neq\emptyset}}\|\Psi_Z\|_{\mathcal{U}}\sum_{z\in\partial_\Psi Z}\sum_{x\in\Lambda}\mathbf{F}(|x-z|)$$

$$\leq\sum_{y\in\Lambda_{L_2}\setminus\Lambda_{L_1}}\sum_{\substack{Z\subseteq\Lambda_{L_2},\ Z\supset\{y\}}}\|\Psi_Z\|_{\mathcal{U}}\sum_{z\in Z}\sum_{x\in\Lambda}\mathbf{F}(|x-z|)$$

$$\leq\sum_{y\in\Lambda_{L_2}\setminus\Lambda_{L_1}}\sum_{z\in\Lambda_{L_2}}\sum_{\substack{Z\subseteq\Lambda_{L_2},\ Z\supset\{y,z\}}}\|\Psi_Z\|_{\mathcal{U}}\sum_{x\in\Lambda}\mathbf{F}(|x-z|)$$

$$\leq\|\Psi\|_{\mathcal{W}}\sum_{y\in\Lambda_{L_2}\setminus\Lambda_{L_1}}\sum_{x\in\Lambda}\sum_{z\in\Lambda_{L_2}}\mathbf{F}(|y-z|)\mathbf{F}(|x-z|)$$

$$\leq\mathbf{D}\|\Psi\|_{\mathcal{W}}\sum_{y\in\Lambda_{L_2}\setminus\Lambda_{L_1}}\sum_{x\in\Lambda}\mathbf{F}(|x-y|)\,, \tag{4.25}$$

while, by using (4.8) only,

$$\sum_{\substack{Z\subseteq\Lambda_{L_2},\ Z\cap(\Lambda_{L_2}\setminus\Lambda_{L_1})\neq\emptyset,\ Z\cap\Lambda\neq\emptyset}}\|\Psi_Z\|_{\mathcal{U}}$$

$$\leq\sum_{y\in\Lambda_{L_2}\setminus\Lambda_{L_1}}\sum_{x\in\Lambda}\sum_{\substack{Z\subseteq\Lambda_{L_2},\ Z\supset\{x,y\}}}\|\Psi_Z\|_{\mathcal{U}}$$

$$\leq\|\Psi\|_{\mathcal{W}}\sum_{y\in\Lambda_{L_2}\setminus\Lambda_{L_1}}\sum_{x\in\Lambda}\mathbf{F}(|x-y|)\,. \tag{4.26}$$

The lemma is then a direct consequence of (4.23)–(4.24) combined with the upper bounds (4.25)–(4.26). ∎

The infinite–volume dynamics is obtained from Lemma 4.4 and the completeness of $\mathcal{U}$. Indeed, from the above lemma, for all $t\in\mathbb{R}$, $\tau_t^{(L)}$ converges strongly on $\mathcal{U}_0$

to τ_t, as $L \to \infty$. By density of $\mathcal{U}_0$ in the Banach space $\mathcal{U}$ and the fact that $\tau_t^{(L)}$ are isometries for all $L \in \mathbb{R}_0^+$ and $t \in \mathbb{R}$, the limit τ_t, $t \in \mathbb{R}$, uniquely defines a $*$–automorphism, also denoted by τ_t, of the C^*–algebra $\mathcal{U}$. $\{\tau_t\}_{t\in\mathbb{R}}$ is clearly a group of $*$–automorphisms on $\mathcal{U}$. Again by the above lemma, for any element B in the dense subset $\mathcal{U}_0 \subset \mathcal{U}$, the convergence of $\tau_t^{(L)}(B)$, as $L \to \infty$, is uniform for t on compacta and $\{\tau_t\}_{t\in\mathbb{R}}$ thus defines a C_0–group on $\mathcal{U}$, that is, a strongly continuous group on $\mathcal{U}$.

We need in the sequel an explicit characterization of the infinitesimal generator of this C_0–group. Since the generator equals (4.3) at finite–volume, one expects that the infinitesimal generator equals on $\mathcal{U}_0$ the linear map δ from $\mathcal{U}_0$ to $\mathcal{U}$ defined by

$$\delta(B) \doteq i \sum_{\Lambda \in \mathcal{P}_f(\mathfrak{L})} [\Psi_\Lambda, B] + i \sum_{x \in \mathfrak{L}} [\mathbf{V}_{\{x\}}, B] , \qquad B \in \mathcal{U}_0 , \qquad (4.27)$$

for any $\Psi \in \mathcal{W}$ and potential $\mathbf{V}$. Indeed, for any $\Lambda \in \mathcal{P}_f(\mathfrak{L})$ and local element $B \in \mathcal{U}_\Lambda$,

$$\sum_{\mathcal{Z} \in \mathcal{P}_f(\mathfrak{L})} \|[\Psi_{\mathcal{Z}}, B]\|_{\mathcal{U}} + \sum_{x \in \mathfrak{L}} \|[\mathbf{V}_{\{x\}}, B]\|_{\mathcal{U}} \qquad (4.28)$$

$$\leq 2 \|B\|_{\mathcal{U}} \left(|\Lambda| \, \mathbf{F}(0) \, \|\Psi\|_{\mathcal{W}} + \sum_{x \in \Lambda} \|\mathbf{V}_{\{x\}}\|_{\mathcal{U}} \right)$$

and the series (4.27) is absolutely convergent for all $B \in \mathcal{U}_0$. Moreover, by (4.3), we obviously have

$$\delta(B) = \lim_{L\to\infty} \delta^{(L)}(B) , \qquad B \in \mathcal{U}_0 . \qquad (4.29)$$

To prove that the closure of the linear map $\delta : \mathcal{U}_0 \to \mathcal{U}$ is the generator of the C_0–group $\{\tau_t\}_{t\in\mathbb{R}}$ of $*$–automorphisms we use the second Trotter–Kato approximation theorem [EN, Chap. III, Sect. 4.9].

To this end, we first show that the (generally unbounded) operator δ on $\mathcal{U}$ with dense domain $\mathrm{Dom}(\delta) = \mathcal{U}_0$ is closable. Observe that both $\pm\delta$ are symmetric derivations and δ is thus conservative [BR1, Definition 3.1.13.], by structure of the set $\mathcal{U}_0$ of local elements:

Lemma 4.5 (Conservative infinite–volume derivation)
Let $\Psi \in \mathcal{W}$ and $\mathbf{V}$ be any potential. Then, the derivation δ defined on $\mathcal{U}_0$ by (4.27) is a conservative symmetric derivation.

Proof Let $B \in \mathcal{U}_0$ satisfying $B \geq 0$. By definition of $\mathcal{U}_0$, $B \in \mathcal{U}_\Lambda$ for some $\Lambda \in \mathcal{P}_f(\mathfrak{L})$. Since $\mathcal{U}_\Lambda$ is a unital C^* –algebra, there is $B^{1/2} \in \mathcal{U}_\Lambda \subset \mathcal{U}_0$ such that $B^{1/2} \geq 0$ and $(B^{1/2})^2 = B$. Therefore, the lemma follows from [BR1, Proposition 3.2.22]. ∎

It follows that the symmetric derivation δ is (norm–) closable:

Lemma 4.6 (Closure of the infinite–volume derivation)
Let $\Psi \in \mathcal{W}$ and $\mathbf{V}$ be any potential. Then, the derivations $\pm\delta$ defined on $\mathcal{U}_0$ by (4.27) are closable and their closures, again denoted for simplicity by $\pm\delta$, are conservative.

Proof $\pm\delta$ are densely defined dissipative operators on the Banach space $\mathcal{U}$. Therefore, the lemma is an obvious application of [BR1, Proposition 3.1.15.]. ∎

In order to apply the second Trotter–Kato approximation theorem [EN, Chap. III, Sect. 4.9], we also prove that the range $\mathrm{Ran}\{(x\mathbf{1}_\mathcal{U} \mp \delta)\}$ of the closed operators $x\mathbf{1}_\mathcal{U} \mp \delta$ are dense in the Banach space $\mathcal{U}$ for $x > 0$. This is done in the following lemma:

Lemma 4.7 (Range of the infinite–volume derivation)
Let $\Psi \in \mathcal{W}$ and $\mathbf{V}$ be any potential. Then, for any $x \in \mathbb{R}^+$,

$$\mathcal{U}_0 \subseteq \mathrm{Ran}\{(x\mathbf{1}_\mathcal{U} \mp \delta)\} \subseteq \mathcal{U}$$

with $\mathbf{1}_\mathcal{U}$ being the identity on $\mathcal{U}$. In particular, $\mathrm{Ran}\{(x\mathbf{1}_\mathcal{U} \mp \delta)\}$ is dense in $\mathcal{U}$.

Proof We only give the proof for the range of the operator $x\mathbf{1}_\mathcal{U} - \delta$, since the other case uses similar arguments.

Note that $\|\tau_t^{(L)}\|_{\mathcal{B}(\mathcal{U})} = 1$ for any $L \in \mathbb{R}_0^+$ and $t \in \mathbb{R}$. Here, $\mathcal{B}(\mathcal{U})$ is the Banach space of bounded linear operators acting on $\mathcal{U}$. Thus, for any $L \in \mathbb{R}_0^+$, $x \in \mathbb{R}^+$, and $B \in \mathcal{U}$, the improper Riemann integral

$$\int_0^\infty e^{-xs} \tau_s^{(L)}(B)\, \mathrm{d}s \doteq \lim_{t \to \infty} \int_0^t e^{-xs} \tau_s^{(L)}(B)\, \mathrm{d}s$$

exists. By [EN, Chap. II, Sect. 1.10], it follows that, for any $L \in \mathbb{R}_0^+$ and $x \in \mathbb{R}^+$, the resolvent $(x\mathbf{1}_\mathcal{U} - \delta^{(L)})^{-1}$ of the generator $\delta^{(L)}$ of the group $\{\tau_t^{(L)}\}_{t \in \mathbb{R}}$ also exists and satisfies

$$(x\mathbf{1}_\mathcal{U} - \delta^{(L)})^{-1}(B) = \int_0^\infty e^{-xs} \tau_s^{(L)}(B)\, \mathrm{d}s \tag{4.30}$$

for all $B \in \mathcal{U}$. Now, take $B \in \mathcal{U}_0$, $x \in \mathbb{R}^+$, and consider the element

$$B_L \doteq (x\mathbf{1}_\mathcal{U} - \delta^{(L)})^{-1}(B) \in \mathcal{U} \tag{4.31}$$

for some sufficiently large parameter $L \in \mathbb{R}_0^+$ such that $B \in \mathcal{U}_{\Lambda_L}$. Note that $\tau_s^{(L)}(\mathcal{U}_{\Lambda_L}) \subset \mathcal{U}_{\Lambda_L}$ and $B_L \in \mathcal{U}_{\Lambda_L} \subset \mathcal{U}_0$ because of (4.30). Then, we observe that

$$(x\mathbf{1}_\mathcal{U} - \delta)(B_L) = B + (\delta^{(L)} - \delta)(B_L)\,,$$

where we recall that $L \in \mathbb{R}_0^+$, $x \in \mathbb{R}^+$, and $B \in \mathcal{U}_0$. Now, by the Lumer–Phillips theorem [BR1, Theorem 3.1.16] (see also its proof), if there is $x \in \mathbb{R}^+$ such that

$$\lim_{L \to \infty} \left\| (\delta - \delta^{(L)})(B_L) \right\|_\mathcal{U} = 0 \tag{4.32}$$

for all $B \in \mathcal{U}_0$ then we obtain the assertion. Indeed, by using Lemma 4.4 together with $\|\tau_t^{(L)}\|_{\mathcal{B}(\mathcal{U})} = 1$ and (4.30), one verifies that $\{B_L\}_{L \in \mathbb{R}_0^+}$ is a Cauchy net, thus a convergent one in $\mathcal{U}$, while $x\mathbf{1}_{\mathcal{U}} - \delta$ is a closed operator, by Lemma 4.6.

To prove (4.32) we use Lieb–Robinson bounds (Theorem 4.3) as follows: Since $B_L \in \mathcal{U}_{\Lambda_L}$ for sufficiently large $L \in \mathbb{R}_0^+$, we can combine (4.3) and (4.27) with (4.30)–(4.31) to compute that

$$(\delta - \delta^{(L)})(B_L) = i \sum_{\mathcal{Z} \in \mathcal{P}_f(\mathfrak{L}),\ \mathcal{Z} \cap \Lambda_L^c \neq \emptyset} \int_0^\infty e^{-xs} \left[\Psi_{\mathcal{Z}}, \tau_s^{(L)}(B)\right] ds \qquad (4.33)$$

for any $x \in \mathbb{R}^+$, sufficiently large $L \in \mathbb{R}_0^+$, and $B \in \mathcal{U}_0$. Here, $\Lambda_L^c \doteq \mathfrak{L} \backslash \Lambda_L$. It suffices to consider the case $B \neq 0$. Using now Theorem 4.3, similar to (4.24), one gets that, for all $s \in \mathbb{R}^+$ and any sufficiently large $L \in \mathbb{R}_0^+$ such that $B \in \mathcal{U}_\Lambda \subset \mathcal{U}_{\Lambda_L}$ with $\Lambda \in \mathcal{P}_f(\mathfrak{L})$,

$$\sum_{\mathcal{Z} \in \mathcal{P}_f(\mathfrak{L}),\ \mathcal{Z} \cap \Lambda_L^c \neq \emptyset} \frac{\left\|\left[\Psi_{\mathcal{Z}}, \tau_s^{(L)}(B)\right]\right\|_{\mathcal{U}}}{2\|B\|_{\mathcal{U}}} \qquad (4.34)$$

$$\leq \mathbf{D}^{-1}\left(e^{2\mathbf{D}|s|\|\Psi\|_{\mathcal{W}}} - 1\right) \sum_{\mathcal{Z} \in \mathcal{P}_f(\mathfrak{L}),\ \mathcal{Z} \cap \Lambda_L^c \neq \emptyset,\ \mathcal{Z} \cap \Lambda = \emptyset} \|\Psi_{\mathcal{Z}}\|_{\mathcal{U}} \sum_{x \in \partial_\Psi \mathcal{Z}} \sum_{y \in \Lambda} \mathbf{F}(|x - y|)$$

$$+ \sum_{\mathcal{Z} \in \mathcal{P}_f(\mathfrak{L}),\ \mathcal{Z} \cap \Lambda_L^c \neq \emptyset,\ \mathcal{Z} \cap \Lambda \neq \emptyset} \|\Psi_{\mathcal{Z}}\|_{\mathcal{U}}\ .$$

Similar to Inequalities (4.25)–(4.26), we thus infer from (4.6) and (4.8) that

$$\sum_{\mathcal{Z} \in \mathcal{P}_f(\mathfrak{L}),\ \mathcal{Z} \cap \Lambda_L^c \neq \emptyset} \frac{\left\|\left[\Psi_{\mathcal{Z}}, \tau_s^{(L)}(B)\right]\right\|_{\mathcal{U}}}{2\|B\|_{\mathcal{U}}} \leq \|\Psi\|_{\mathcal{W}}\, e^{2\mathbf{D}|s|\|\Psi\|_{\mathcal{W}}} \sum_{y \in \Lambda_L^c} \sum_{x \in \Lambda} \mathbf{F}(|x - y|)\ , \qquad (4.35)$$

while

$$\lim_{L \to \infty} \sum_{y \in \Lambda_L^c} \sum_{x \in \Lambda} \mathbf{F}(|x - y|) = 0\ , \qquad (4.36)$$

because of (4.5). Therefore, by (4.33)–(4.36), we deduce (4.32) for all $x > 2\mathbf{D}\|\Psi\|_{\mathcal{W}}$ and $B \in \mathcal{U}_0$. ∎

We now apply the second Trotter–Kato approximation theorem [EN, Chap. III, Sect. 4.9] to deduce that δ is the generator of the group $\{\tau_t\}_{t \in \mathbb{R}}$ of $*$–automorphisms and resume all the main results, so far, in the following theorem:

Theorem 4.8 (Infinite–volume dynamics and its generator)
Let $\Psi \in \mathcal{W}$, $\mathbf{V}$ be any potential, and $\mathbf{D} \in \mathbb{R}^+$ be defined by (4.6).
(i) Infinite–volume dynamics. The continuous groups $\{\tau_t^{(L)}\}_{t \in \mathbb{R}}$, $L \in \mathbb{R}_0^+$, defined by (4.2) converge strongly to a C_0–group $\{\tau_t\}_{t \in \mathbb{R}}$ of $*$–automorphisms with generator δ.

(ii) *Infinitesimal generator.* δ *is a conservative closed symmetric derivation which is equal on its core* $\mathcal{U}_0$ *to*

$$\delta(B) = i \sum_{\Lambda \in \mathcal{P}_f(\mathfrak{L})} [\Psi_\Lambda, B] + i \sum_{x \in \mathfrak{L}} [\mathbf{V}_{\{x\}}, B] , \qquad B \in \mathcal{U}_0 .$$

(iii) *Rate of convergence. For any* $\Lambda \in \mathcal{P}_f(\mathfrak{L})$, $B \in \mathcal{U}_\Lambda$ *and* $L \in \mathbb{R}_0^+$ *such that* $\Lambda \subset \Lambda_L$,

$$\left\| \tau_t(B) - \tau_t^{(L)}(B) \right\|_{\mathcal{U}} \leq 2 \|B\|_{\mathcal{U}} \|\Psi\|_{\mathcal{W}} |t| e^{4\mathbf{D}|t| \|\Psi\|_{\mathcal{W}}} \sum_{y \in \mathfrak{L} \backslash \Lambda_L} \sum_{x \in \Lambda} \mathbf{F}(|x - y|) .$$

(iv) *Lieb–Robinson bounds. For any* $t \in \mathbb{R}$ *and* $B_1 \in \mathcal{U}^+ \cap \mathcal{U}_{\Lambda^{(1)}}$, $B_2 \in \mathcal{U}_{\Lambda^{(2)}}$ *with disjoint sets* $\Lambda^{(1)}, \Lambda^{(2)} \in \mathcal{P}_f(\mathfrak{L})$,

$$\|[\tau_t(B_1), B_2]\|_{\mathcal{U}} \leq 2\mathbf{D}^{-1} \|B_1\|_{\mathcal{U}} \|B_2\|_{\mathcal{U}} \left(e^{2\mathbf{D}|t| \|\Psi\|_{\mathcal{W}}} - 1 \right)$$
$$\times \sum_{x \in \partial_\Psi \Lambda^{(1)}} \sum_{y \in \Lambda^{(2)}} \mathbf{F}(|x - y|) .$$

Proof By Lemma 4.6, the set $\mathcal{U}_0$ of local elements is a core of the dissipative derivation δ and one obtains (ii), see (4.27). Moreover, $\delta^{(L)}(B) \to \delta(B)$ for all $B \in \mathcal{U}_0$, see (4.29). Recall that $\delta^{(L)}$ is the generator of the group $\{\tau_t^{(L)}\}_{t \in \mathbb{R}}$ for any $L \in \mathbb{R}_0^+$. Therefore, since one also has Lemma 4.7, (i) is a direct consequence of [EN, Chap. III, Sect. 4.9]. The third statement (iii) thus follows from Lemma 4.4. (iv) is an obvious consequence of Theorem 4.3 and the first assertion (i). ∎

4.4 Lieb–Robinson Bounds for Multi-commutators

Recall that multi–commutators are defined by induction as follows:

$$[B_1, B_0]^{(2)} \doteq [B_1, B_0] \doteq B_1 B_0 - B_0 B_1 , \qquad B_0, B_1 \in \mathcal{U} , \qquad (4.37)$$

and, for all integers $k \geq 2$,

$$[B_k, B_{k-1}, \ldots, B_0]^{(k+1)} \doteq [B_k, [B_{k-1}, \ldots, B_0]^{(k)}] , \qquad B_0, \ldots, B_k \in \mathcal{U} . \qquad (4.38)$$

The aim of this subsection is to extend Theorem 4.8 (iv) to multi–commutators. The arguments we use below to prove Lieb–Robinson bounds for multi–commutators are not a generalization of the proof of Theorems 4.3 or 4.8 (iv). Instead, we use a pivotal lemma deduced from Theorem 4.8 (iii), which in turn results from finite–volume Lieb–Robinson bounds of Theorem 4.3. This lemma expresses the C_0–group $\{\tau_t\}_{t \in \mathbb{R}}$ of Theorem 4.8 (i) as *telescoping* series.

To this end, it is convenient to introduce the family $\{\chi_x\}_{x\in\mathfrak{L}}$ of $*$–automorphisms of $\mathcal{U}$, which implements the action of the group of lattice translations on the CAR C^*–algebra $\mathcal{U}$. This family is uniquely defined by the conditions

$$\chi_x(a_y) = a_{y+x} , \qquad x, y \in \mathfrak{L} . \tag{4.39}$$

We also define, for any $n \in \mathbb{N}_0$, $x \in \mathfrak{L}$, $\Psi \in \mathcal{W}$ and potential $\mathbf{V}$, a *space trans-lated* finite–volume dynamics which is the continuous group $\{\tau_t^{(n,x)}\}_{t\in\mathbb{R}}$ of $*$–automorphisms of $\mathcal{U}$ generated by the symmetric and bounded derivation

$$\delta^{(n,x)}(B) \doteq i \sum_{\Lambda\subseteq x+\Lambda_n} [\Psi_\Lambda, B] + i \sum_{y\in x+\Lambda_n} \left[\mathbf{V}_{\{y\}}, B\right] , \qquad B \in \mathcal{U} .$$

Note that the fermion system is generally *not* translation invariant and, in general,

$$\tau_t^{(n,x)} \circ \chi_x \neq \chi_x \circ \tau_t^{(n)} , \qquad x \in \mathfrak{L}, n \in \mathbb{N}_0 , t \in \mathbb{R} .$$

For $m \in \mathbb{N}_0$, $x \in \mathfrak{L}$, $B \in \mathcal{U}_{\Lambda_m}$ and $t \in \mathbb{R}$, we finally introduce the local elements

$$\mathfrak{B}_{B,t,x}(m) \equiv \mathfrak{B}_{B,t,x}^{(m)}(m) \doteq \tau_t^{(m,x)} \circ \chi_x(B) \in \mathcal{U}_{\Lambda_m+x} \tag{4.40}$$

and

$$\mathfrak{B}_{B,t,x}(n) \equiv \mathfrak{B}_{B,t,x}^{(m)}(n) \doteq (\tau_t^{(n,x)} - \tau_t^{(n-1,x)}) \circ \chi_x(B) \in \mathcal{U}_{\Lambda_n+x} , \qquad n \geq m+1 . \tag{4.41}$$

The family $\{\mathfrak{B}_{B,t,x}(n)\}_{n\geq m} \subset \mathcal{U}_0$ is used to define telescoping series:

Lemma 4.9 (Infinite–volume dynamics as telescoping series)
Let $\Psi \in \mathcal{W}$ and $\mathbf{V}$ be any potential. Then, for any $m \in \mathbb{N}_0$, $x \in \mathfrak{L}$, $B \in \mathcal{U}_{\Lambda_m}$ and $t \in \mathbb{R}$:

$$\sum_{n=m}^{\infty} \mathfrak{B}_{B,t,x}(n) = \tau_t \circ \chi_x(B) . \tag{4.42}$$

The above telescoping series is absolutely convergent in $\mathcal{U}$ with

$$\left\|\mathfrak{B}_{B,t,x}(n)\right\|_{\mathcal{U}} \leq 2 \|B\|_{\mathcal{U}} \|\Psi\|_{\mathcal{W}} |t| e^{4\mathbf{D}|t|\|\Psi\|_{\mathcal{W}}} \sum_{y\in\Lambda_n\backslash\Lambda_{n-1}} \sum_{z\in\Lambda_m} \mathbf{F}(|z-y|) \tag{4.43}$$

for any $n \geq m+1$, while $\left\|\mathfrak{B}_{B,t,x}(m)\right\|_{\mathcal{U}} = \|B\|_{\mathcal{U}}$.

Proof Since, for any $N \in \mathbb{N}_0$ so that $N \geq m$,

$$\sum_{n=m}^{N} \mathfrak{B}_{B,t,x}(n) = \tau_t^{(N,x)} \circ \chi_x(B) , \tag{4.44}$$

it suffices to study the limit $N \to \infty$ of the group $\{\tau_t^{(N,x)}\}_{t\in\mathbb{R}}$ at any fixed $x \in \mathfrak{L}$. Similar to the proof of Theorem 4.8 (i), $\delta^{(N,x)}(B) \to \delta(B)$ for all $B \in \mathcal{U}_0$, as $N \to \infty$. By Lemma 4.7 and [EN, Chap. III, Sect. 4.9], the translated groups $\{\tau_t^{(N,x)}\}_{t\in\mathbb{R}}$, $N \in \mathbb{N}_0$, converge strongly to the C_0–group $\{\tau_t\}_{t\in\mathbb{R}}$ for any $x \in \mathfrak{L}$. In other words, we deduce Eq. (4.42) from (4.44) in the limit $N \to \infty$. Moreover, one easily checks that Theorem 4.3 and thus Lemma 4.4 also hold for the (space translated) groups $\{\tau_t^{(n,x)}\}_{t\in\mathbb{R}}, n \in \mathbb{N}_0$, at any fixed $x \in \mathfrak{L}$. This yields Inequality (4.43) for $n > m$, while $\left\| \mathfrak{B}_{B,t,x}(m) \right\|_{\mathcal{U}} = \|B\|_{\mathcal{U}}$, because $\tau_t^{(m,x)}$ is a $*$–automorphism on $\mathcal{U}_{\Lambda_m}$. It follows that

$$\sum_{n=m+1}^{\infty} \left\| \mathfrak{B}_{B,t,x}(n) \right\|_{\mathcal{U}} \leq 2\|B\|_{\mathcal{U}} \|\Psi\|_{\mathcal{W}} |t| \, e^{4\mathbf{D}|t|\|\Psi\|_{\mathcal{W}}} \sum_{z\in\Lambda_m} \sum_{n\in\mathbb{N}} \sum_{y\in\Lambda_n\setminus\Lambda_{n-1}} \mathbf{F}(|z-y|) \, .$$

Finally, by Assumption (4.5),

$$\sum_{z\in\Lambda_m} \sum_{n\in\mathbb{N}} \sum_{y\in\Lambda_n\setminus\Lambda_{n-1}} \mathbf{F}(|z-y|) \leq \sum_{z\in\Lambda_m} \sum_{y\in\mathfrak{L}} \mathbf{F}(|z-y|) = |\Lambda_m| \, \|\mathbf{F}\|_{1,\mathfrak{L}} < \infty \, .$$

■

To extend Lieb–Robinson bounds to multi–commutators we combine Lemma 4.9 with tree decompositions of sequences of clustering subsets of $\mathfrak{L}$ (cf. (4.53)): Let $\mathcal{T}_2$ be the set of all (non–oriented) trees with exactly two vertices. This set contains a unique tree $T = \{\{0, 1\}\}$ which, in turn, contains the unique bond $\{0, 1\}$, i.e., $\mathcal{T}_2 \doteq \{\{\{0, 1\}\}\}$. Then, for each integer $k \geq 2$, we recursively define a set $\mathcal{T}_{k+1}$ of trees with $k+1$ vertices by

$$\mathcal{T}_{k+1} \doteq \left\{ \{\{j, k\}\} \cup T \; : \; j = 0, \ldots, k-1, \quad T \in \mathcal{T}_k \right\} . \tag{4.45}$$

Therefore, for $k \in \mathbb{N}$ and any tree $T \in \mathcal{T}_{k+1}$, there is a map

$$\mathrm{P}_T : \{1, \ldots, k\} \to \{0, \ldots, k-1\} \tag{4.46}$$

such that $\mathrm{P}_T(j) < j, \mathrm{P}_T(1) = 0$, and

$$T = \bigcup_{j=1}^{k} \{\{\mathrm{P}_T(j), j\}\} . \tag{4.47}$$

For any $k \in \mathbb{N}$, $T \in \mathcal{T}_{k+1}$, and every sequence $\{(n_j, x_j)\}_{j=0}^{k}$ in $\mathbb{N}_0 \times \mathfrak{L}$ with length $k+1$, we define

$$\varkappa_T\left(\{(n_j, x_j)\}_{j=0}^{k}\right) \doteq \prod_{j=1}^{k} \mathbf{1}\left[(\Lambda_{n_j} + x_j) \cap (\Lambda_{n_{\mathrm{P}_T(j)}} + x_{\mathrm{P}_T(j)}) \neq \emptyset\right] \in \{0, 1\} ,$$

$$\tag{4.48}$$

while, for all $\ell \in \{1, \ldots, k\}$,

$$\mathcal{S}_{\ell,k} \doteq \{\pi \mid \pi : \{\ell, \ldots, k\} \to \{1, \ldots, k\} \text{ such that } \pi\,(i) < \pi\,(j) \text{ when } i < j\}\,. \tag{4.49}$$

Then, one gets the following bound on multi–commutators:

Theorem 4.10 (Lieb–Robinson bounds for multi–commutators – Part I)
Let $\Psi \in \mathcal{W}$ *and* $\mathbf{V}$ *be any potential. Then, for any integer* $k \in \mathbb{N}$, $\{m_j\}_{j=0}^k \subset \mathbb{N}_0$, *times* $\{s_j\}_{j=1}^k \subset \mathbb{R}$, *lattice sites* $\{x_j\}_{j=0}^k \subset \mathfrak{L}$, *and local elements* $B_0 \in \mathcal{U}_0$, $\{B_j\}_{j=1}^k \subset \mathcal{U}_0 \cap \mathcal{U}^+$ *such that* $B_j \in \mathcal{U}_{\Lambda_{m_j}}$ *for* $j \in \{0, \ldots, k\}$,

$$\left\| \left[\tau_{s_k} \circ \chi_{x_k}(B_k), \ldots, \tau_{s_1} \circ \chi_{x_1}(B_1), \chi_{x_0}(B_0) \right]^{(k+1)} \right\|_{\mathcal{U}}$$

$$\leq 2^k \prod_{j=0}^k \|B_j\|_{\mathcal{U}} \sum_{T \in \mathcal{T}_{k+1}} \left(\varkappa_T \left(\{(m_j, x_j)\}_{j=0}^k \right) + \Re_{T, \|\Psi\|_{\mathcal{W}}} \right)$$

with, for any $\alpha \in \mathbb{R}_0^+$,

$$\Re_{T,\alpha} \doteq \sum_{\ell=1}^k (2\alpha)^{k-\ell+1} \sum_{\pi \in \mathcal{S}_{\ell,k}} \left(\prod_{j \in \{\pi(\ell), \ldots, \pi(k)\}} |s_j| \, e^{4\mathbf{D}\alpha|s_j|} \right) \tag{4.50}$$

$$\sum_{n_{\pi(\ell)}=m_{\pi(\ell)}+1}^{\infty} \sum_{z_{\pi(\ell)} \in \Lambda_{m_{\pi(\ell)}}} \sum_{y_{\pi(\ell)} \in \Lambda_{n_{\pi(\ell)}} \setminus \Lambda_{n_{\pi(\ell)}-1}} \cdots$$

$$\cdots \sum_{n_{\pi(k)}=m_{\pi(k)}+1}^{\infty} \sum_{z_{\pi(k)} \in \Lambda_{m_{\pi(k)}}} \sum_{y_{\pi(k)} \in \Lambda_{n_{\pi(k)}} \setminus \Lambda_{n_{\pi(k)}-1}}$$

$$\varkappa_T \left(\{(n_j, x_j)\}_{j=0}^k \right) \prod_{j \in \{\pi(\ell), \ldots, \pi(k)\}} \mathbf{F}\left(|z_j - y_j| \right)\,.$$

In the right–hand side (r.h.s.) of (4.50), we set $n_j \doteq m_j$ *if*

$$j \in \{0, \ldots, k\} \setminus \{\pi\,(\ell), \ldots, \pi\,(k)\}\,.$$

The constant $\mathbf{D} \in \mathbb{R}^+$ *is defined by (4.6).*

Proof Fix $k \in \mathbb{N}$, $\{m_j\}_{j=0}^k \subset \mathbb{N}_0$, $\{s_j\}_{j=1}^k \subset \mathbb{R}$, $\{x_j\}_{j=0}^k \subset \mathfrak{L}$ and elements $\{B_j\}_{j=0}^k \subset \mathcal{U}_0$ such that the conditions of the theorem are satisfied. From Lemma 4.9,

$$\left[\tau_{s_k} \circ \chi_{x_k}(B_k), \ldots, \tau_{s_1} \circ \chi_{x_1}(B_1), \chi_{x_0}(B_0) \right]^{(k+1)} \tag{4.51}$$

$$= \sum_{n_1=m_1}^{\infty} \cdots \sum_{n_k=m_k}^{\infty} \left[\mathfrak{B}_{B_k,s_k,x_k}(n_k), \ldots, \mathfrak{B}_{B_1,s_1,x_1}(n_1), \chi_{x_0}(B_0) \right]^{(k+1)}\,.$$

Since $B_j \in \mathcal{U}_{\Lambda_{m_j}} \cap \mathcal{U}^+$ for $j \in \{1, \ldots, k\}$, we infer from (4.40)–(4.41) that

$$
\left[\mathfrak{B}_{B_k, s_k, x_k}(n_k), \ldots, \mathfrak{B}_{B_1, s_1, x_1}(n_1), \chi_{x_0}(B_0) \right]^{(k+1)}
$$

$$
= \prod_{j=1}^{k} \mathbf{1} \left[\bigcup_{i=0}^{j-1} (\Lambda_{n_j} + x_j) \cap (\Lambda_{n_i} + x_i) \neq \emptyset \right] \tag{4.52}
$$

$$
\left[\mathfrak{B}_{B_k, s_k, x_k}(n_k), \ldots, \mathfrak{B}_{B_1, s_1, x_1}(n_1), \chi_{x_0}(B_0) \right]^{(k+1)}
$$

for all integers $\{n_j\}_{j=0}^{k} \subset \mathbb{N}_0$ with $n_0 \doteq m_0$ and $n_j \geq m_j$ when $j \in \{1, \ldots, k\}$. The conditions inside characteristic functions in (4.52) refer to the fact that the sequence of sets $\{\Lambda_{n_j}\}_{j=0}^{k}$ has to be a cluster to have a non–zero multi–commutator. Note further that

$$
\prod_{j=1}^{k} \mathbf{1} \left[\bigcup_{i=0}^{j-1} (\Lambda_{n_j} + x_j) \cap (\Lambda_{n_i} + x_i) \neq \emptyset \right] \leq \sum_{T \in \mathcal{T}_{k+1}} \varkappa_T \left(\{(n_j, x_j)\}_{j=0}^{k} \right). \tag{4.53}
$$

Using (4.51)–(4.53) one then shows that

$$
\left\| \left[\tau_{s_k} \circ \chi_{x_k}(B_k), \ldots, \tau_{s_1} \circ \chi_{x_1}(B_1), \chi_{x_0}(B_0) \right]^{(k+1)} \right\|_{\mathcal{U}}
$$

$$
\leq 2^k \|B_0\|_{\mathcal{U}} \sum_{T \in \mathcal{T}_{k+1}} \sum_{n_1=m_1}^{\infty} \cdots \sum_{n_k=m_k}^{\infty} \varkappa_T \left(\{(n_j, x_j)\}_{j=0}^{k} \right)
$$

$$
\times \prod_{j=1}^{k} \left\| \mathfrak{B}_{B_j, s_j, x_j}(n_j) \right\|_{\mathcal{U}}. \tag{4.54}
$$

This inequality combined with (4.43) yields the assertion. ∎

The above theorem extends Lieb–Robinson bounds to multi–commutators. Indeed, if $\mathbf{F}(r)$ decays fast enough as $r \to \infty$, then Theorem 4.10 and Lebesgue's dominated convergence theorem imply that, for any $j \in \{0, \ldots, k\}$,

$$
\lim_{|x_j| \to \infty} \left\| \left[\tau_{s_k} \circ \chi_{x_k}(B_k), \ldots, \tau_{s_1} \circ \chi_{x_1}(B_1), \chi_{x_0}(B_0) \right]^{(k+1)} \right\|_{\mathcal{U}} = 0. \tag{4.55}
$$

The rate of convergence if this multi–commutator towards zero is, however, a priori unclear. Hence, to obtain bounds on the space decay of the above multi–commutator, more in the spirit of the original Lieb–Robinson bounds for commutators, we consider two situations w.r.t. the behavior of the function $\mathbf{F} : \mathbb{R}_0^+ \to \mathbb{R}^+$ at large arguments:

- *Polynomial decay.* There is a constant $\varsigma \in \mathbb{R}^+$ and, for all $m \in \mathbb{N}_0$, an absolutely summable sequence $\{\mathbf{u}_{n,m}\}_{n \in \mathbb{N}} \in \ell^1(\mathbb{N})$ such that, for all $n \in \mathbb{N}$ with $n > m$,

$$|\Lambda_n \backslash \Lambda_{n-1}| \sum_{z \in \Lambda_m} \max_{y \in \Lambda_n \backslash \Lambda_{n-1}} \mathbf{F}\left(|z - y|\right) \leq \frac{\mathbf{u}_{n,m}}{(1+n)^s} . \tag{4.56}$$

- *Exponential decay.* There is $\varsigma \in \mathbb{R}^+$ and, for $m \in \mathbb{N}_0$, a constant $\mathbf{C}_m \in \mathbb{R}^+$ such that, for all $n \in \mathbb{N}$ with $n > m$,

$$|\Lambda_n \backslash \Lambda_{n-1}| \sum_{z \in \Lambda_m} \max_{y \in \Lambda_n \backslash \Lambda_{n-1}} \mathbf{F}\left(|z - y|\right) \leq \mathbf{C}_m e^{-2\varsigma n} . \tag{4.57}$$

For sufficiently large $\epsilon \in \mathbb{R}^+$, the function (4.7) clearly satisfies Condition (4.56), while (4.56)–(4.57) hold for the choice

$$\mathbf{F}(r) = e^{-2\varsigma r}(1+r)^{-(d+\epsilon)} , \qquad r \in \mathbb{R}_0^+ , \tag{4.58}$$

with arbitrary $\varsigma, \epsilon \in \mathbb{R}^+$. Under one of these both very general assumptions, one can put the upper bound of Theorem 4.10 in a much more convenient form. In fact, one obtains an estimate on the norm of the multi–commutator (4.55) as a function of the distances between the points $\{x_0, \ldots, x_k\}$, like in the usual Lieb–Robinson bounds (i.e., the special case $k = 2$). To formulate such bounds, we need some preliminary definitions related to properties of trees.

For any $k \in \mathbb{N}$ and $T \in \mathcal{T}_{k+1}$, we define the sequence $\partial_T \equiv \{\partial_T(j)\}_{j=0}^k$ in $\{1, \ldots, k\}$ by

$$\partial_T(j) \doteq |\{b \in T : j \in b\}| , \qquad j \in \{0, \ldots, k\} ,$$

i.e., $\partial_T(j)$ is the *degree* of the j-th vertex of the tree T. For $k \in \mathbb{N}$ and $T \in \mathcal{T}_{k+1}$, observe that

$$\partial_T(0) + \cdots + \partial_T(k) = 2k . \tag{4.59}$$

We also introduce the following notation:

$$\partial_T! \doteq \partial_T(0)! \cdots \partial_T(k)!$$

for any tree $T \in \mathcal{T}_{k+1}$, $k \in \mathbb{N}$. The degree of any vertex of a tree is at least 1, by connectedness of such a graph, and (4.59) yields

$$\partial_T! \leq k! , \qquad k \in \mathbb{N} , \ T \in \mathcal{T}_{k+1} . \tag{4.60}$$

For any $k \in \mathbb{N}$, $T \in \mathcal{T}_{k+1}$, and any sequence $f : \mathbb{N}_0 \to \mathbb{R}^+$, note that

$$\prod_{j=0}^k \{f(j)\}^{\partial_T(j)} = \prod_{j=1}^k f(j) f(\mathrm{P}_T(j)) . \tag{4.61}$$

This property is elementary but pivotal to estimate the remainder $\mathfrak{R}_{T,\alpha}$, defined by (4.50), of Theorem 4.10.

Theorem 4.11 (Lieb–Robinson bounds for multi–commutators – Part II)
Let $\alpha \in \mathbb{R}_0^+$, $k \in \mathbb{N}$, $\{m_j\}_{j=0}^k \subset \mathbb{N}_0$, $\{s_j\}_{j=1}^k \subset \mathbb{R}$, $\{x_j\}_{j=0}^k \subset \mathcal{L}$, and $T \in \mathcal{T}_{k+1}$. Depending on decay properties of the function $\mathbf{F} : \mathbb{R}_0^+ \to \mathbb{R}^+$, the coefficient $\mathfrak{R}_{T,\alpha} \in \mathbb{R}_0^+$ defined by (4.50) satisfies the following bounds:
(i) *Polynomial decay: Assume (4.56). Then,*

$$
\mathfrak{R}_{T,\alpha} \leq d^{\frac{\varsigma k}{2}} \sum_{\ell=1}^k (2\alpha)^{k-\ell+1} \sum_{\pi \in \mathcal{S}_{\ell,k}} \left(\prod_{j \in \{\pi(\ell),\ldots,\pi(k)\}} \|\mathbf{u}_{\cdot,m_j}\|_{\ell^1(\mathbb{N})} |s_j| \, e^{4\mathbf{D}|s_j|\alpha} \right)
$$
$$
\left(\prod_{j \in \{0,\ldots,k\} \setminus \{\pi(\ell),\ldots,\pi(k)\}} (1 + m_j)^\varsigma \right) \prod_{\{j,l\} \in T} \frac{1}{(1 + |x_j - x_l|)^{\varsigma(\max\{\partial_T(j), \partial_T(l)\})^{-1}}} \, .
$$

(ii) *Exponential decay: Assume (4.57). Then,*

$$
\mathfrak{R}_{T,\alpha} \leq \sum_{\ell=1}^k \left(\frac{2\alpha}{e^\varsigma - 1} \right)^{k-\ell+1} \sum_{\pi \in \mathcal{S}_{\ell,k}} \left(\prod_{j \in \{\pi(\ell),\ldots,\pi(k)\}} C_{m_j} |s_j| \, e^{4\mathbf{D}|s_j|\alpha - \varsigma m_j} \right)
$$
$$
\left(\prod_{j \in \{0,\ldots,k\} \setminus \{\pi(\ell),\ldots,\pi(k)\}} e^{\varsigma m_j} \right) \prod_{\{j,l\} \in T} \exp\left(-\frac{\varsigma |x_j - x_l|}{\sqrt{d} \max\{\partial_T(j), \partial_T(l)\}} \right) \, .
$$

Proof (i) Fix all parameters of the theorem. We infer from (4.50) and (4.56) that

$$
\mathfrak{R}_{T,\alpha} \leq \sum_{\ell=1}^k (2\alpha)^{k-\ell+1} \sum_{\pi \in \mathcal{S}_{\ell,k}} \left(\prod_{j \in \{\pi(\ell),\ldots,\pi(k)\}} |s_j| \, e^{4\mathbf{D}|s_j|\alpha} \right) \sum_{n_{\pi(\ell)}=m_{\pi(\ell)}+1}^\infty
$$
$$
\cdots \sum_{n_{\pi(k)}=m_{\pi(k)}+1}^\infty \varkappa_T \left(\{(n_j, x_j)\}_{j=0}^k \right) \prod_{j \in \{\pi(\ell),\ldots,\pi(k)\}} \frac{\mathbf{u}_{n_j,m_j}}{(1 + n_j)^\varsigma} \, .
$$

Recall that $n_j \doteq m_j$ when $j \in \{0, \ldots, k\} \setminus \{\pi(\ell), \ldots, \pi(k)\}$. By Hölder's inequality, it follows that

$$
\mathfrak{R}_{T,\alpha} \leq \sum_{\ell=1}^k (2\alpha)^{k-\ell+1} \sum_{\pi \in \mathcal{S}_{\ell,k}} \left(\prod_{j \in \{\pi(\ell),\ldots,\pi(k)\}} \|\mathbf{u}_{\cdot,m_j}\|_{\ell^1(\mathbb{N})} |s_j| \, e^{4\mathbf{D}|s_j|\alpha} \right) \tag{4.62}
$$
$$
\times \max_{n_{\pi(\ell)},\ldots,n_{\pi(k)} \in \mathbb{N}} \left\{ \varkappa_T \left(\{(n_j, x_j)\}_{j=0}^k \right) \prod_{j \in \{\pi(\ell),\ldots,\pi(k)\}} \frac{1}{(1 + n_j)^\varsigma} \right\} \, .
$$

Therefore, it suffices to bound the above maximum in an appropriate way. Using (4.61), note that

$$\prod_{j=0}^{k} \frac{1}{(1+n_j)^\varsigma} = \prod_{j=0}^{k} \left(\frac{1}{(1+n_j)^{\frac{\varsigma}{\partial_T(j)}}} \right)^{\partial_T(j)}$$

$$= \prod_{j=1}^{k} \frac{1}{(1+n_j)^{\frac{\varsigma}{\partial_T(j)}} (1+n_{\mathrm{P}_T(j)})^{\frac{\varsigma}{\partial_T(\mathrm{P}_T(j))}}}$$

$$\leq \prod_{j=1}^{k} \frac{1}{(1+n_j+n_{\mathrm{P}_T(j)})^{\frac{\varsigma}{\mathfrak{m}_T(j)}}} \,, \tag{4.63}$$

where, for $k \in \mathbb{N}$, any tree $T \in \mathcal{T}_{k+1}$, and $j \in \{1, \ldots, k\}$,

$$\mathfrak{m}_T(j) \doteq \max\{\partial_T(j), \partial_T(\mathrm{P}_T(j))\} \,.$$

Meanwhile, the condition

$$(\Lambda_{n_j} + x_j) \cap (\Lambda_{n_{\mathrm{P}_T(j)}} + x_{\mathrm{P}_T(j)}) \neq \emptyset$$

implies

$$\sqrt{d}(n_j + n_{\mathrm{P}_T(j)}) \geq |x_j - x_{\mathrm{P}_T(j)}| \,. \tag{4.64}$$

Therefore, we infer from (4.63)–(4.64) that

$$\max_{n_{\pi(\ell)}, \ldots, n_{\pi(k)} \in \mathbb{N}} \left\{ \varkappa_T \left(\{(n_j, x_j)\}_{j=0}^{k} \right) \prod_{j \in \{\pi(\ell), \ldots, \pi(k)\}} \frac{1}{(1+n_j)^\varsigma} \right\}$$

$$\leq \left(\prod_{j \in \{0, \ldots, k\} \setminus \{\pi(\ell), \ldots, \pi(k)\}} (1+n_j)^\varsigma \right) \prod_{j=1}^{k} \frac{d^{\frac{\varsigma}{2}}}{(1+|x_j - x_{\mathrm{P}_T(j)}|)^{\frac{\varsigma}{\mathfrak{m}_T(j)}}} \,.$$

Combined with (4.62), this last inequality yields Assertion (i).

(ii) The second assertion is proven exactly in the same way. We omit the details. ∎

We defined in [BPH1, Sect. 4] the concept of *tree–decay bounds* for pairs (ρ, τ), where $\rho \in \mathcal{U}^*$ and $\tau \equiv \{\tau_t\}_{t \in \mathbb{R}}$ are respectively any state and any one–parameter group of *–automorphisms on the C^*–algebra $\mathcal{U}$. They are a useful tool to control multi–commutators of products of annihilation and creation operators. Such bounds are related to cluster or graph expansions in statistical physics. For more details see the preliminary discussions of [BPH1, Sect. 4]. As a straightforward corollary of Theorems 4.10–4.11 we give below an extension of the tree–decay bounds [BPH1, Sect. 4] to the case of interacting fermions on lattices:

Corollary 4.12 (Tree–decay bounds)
Let $\Psi \in \mathcal{W}$, $\mathbf{V}$ be any potential, $k \in \mathbb{N}$, $m_0 \in \mathbb{N}_0$, $t \in \mathbb{R}_0^+$, $\{s_j\}_{j=1}^{k} \subset [-t, t]$, $B_0 \subset \mathcal{U}_{\Lambda_{m_0}}$, and $\{x_j\}_{j=0}^{k}, \{z_j\}_{j=1}^{k} \subset \mathfrak{L}$ such that $|z_j| = 1$ for $j \in \{1, \ldots, k\}$.

(i) *Polynomial decay: Assume (4.56) for* $m = 1$. *Then,*

$$\left\| \left[\tau_{s_k}(a^*_{x_k}a_{x_k+z_k}), \ldots, \tau_{s_1}(a^*_{x_1}a_{x_1+z_1}), \chi_{x_0}(B_0) \right]^{(k+1)} \right\|_{\mathcal{U}}$$

$$\leq \|B_0\|_{\mathcal{U}} (1+m_0)^\varsigma \mathbf{K}_0^k \sum_{T \in \mathcal{T}_{k+1}} \prod_{\{j,l\} \in T} \frac{1}{(1+|x_j - x_l|)^{\varsigma(\max\{\partial_T(j), \partial_T(l)\})^{-1}}}$$

with

$$\mathbf{K}_0 \doteq 2d^{\frac{\varsigma}{2}} \left(2^\varsigma + 2 \|\mathbf{u}_{\cdot,1}\|_{\ell^1(\mathbb{N})} \|\Psi\|_{\mathcal{W}} |t| \, e^{4\mathbf{D}|t| \|\Psi\|_{\mathcal{W}}} \right).$$

(ii) *Exponential decay: Assume (4.57) for* $m = 1$. *Then,*

$$\left\| \left[\tau_{s_k}(a^*_{x_k}a_{x_k+z_k}), \ldots, \tau_{s_1}(a^*_{x_1}a_{x_1+z_1}), \chi_{x_0}(B_0) \right]^{(k+1)} \right\|_{\mathcal{U}}$$

$$\leq \|B_0\|_{\mathcal{U}} \, e^{m_0\varsigma} \mathbf{K}_1^k \sum_{T \in \mathcal{T}_{k+1}} \prod_{\{j,l\} \in T} \exp\left(-\frac{\varsigma |x_j - x_l|}{\sqrt{d} \max\{\partial_T(j), \partial_T(l)\}} \right)$$

with

$$\mathbf{K}_1 \doteq 2 \left(e^\varsigma + \frac{2C_1 \|\Psi\|_{\mathcal{W}} |t| \, e^{4\mathbf{D}|t| \|\Psi\|_{\mathcal{W}}}}{e^{2\varsigma} - e^\varsigma} \right).$$

Proof For all $k \in \mathbb{N}$, $T \in \mathcal{T}_{k+1}$, and any sequence $\{(m_j, x_j)\}_{j=0}^k$ in $\mathbb{N}_0 \times \mathcal{L}$ of length $k + 1$, the following upper bounds hold for $\varkappa_T$ (see (4.48)):

$$\varkappa_T \left(\{(m_j, x_j)\}_{j=0}^k \right) \leq d^{\frac{k\varsigma}{2}} \prod_{j=0}^k (1+m_j)^\varsigma \prod_{\{j,l\} \in T} \frac{1}{(1+|x_j - x_l|)^{\frac{\varsigma}{\max\{\partial_T(j), \partial_T(l)\}}}} \quad (4.65)$$

while

$$\varkappa_T \left(\{(m_j, x_j)\}_{j=0}^k \right) \leq e^{(m_0+\cdots+m_k)\varsigma} \prod_{\{j,l\} \in T} \exp\left(-\frac{\varsigma |x_j - x_l|}{\sqrt{d} \max\{\partial_T(j), \partial_T(l)\}} \right). \quad (4.66)$$

Cf. proof of Theorem 4.11. Therefore, the corollary is a direct consequence of Theorems 4.10 and 4.11 together with the two previous inequalities. ∎

Up to the powers $1/\max\{\partial_T(j), \partial_T(l)\}$, Corollary 4.12 gives for interacting systems upper bounds for multi–commutators like [BPH1, Eq. (4.14)] for the free case. We show in the next subsection how to use these bounds to obtain results similar to [BPH1, Theorem 3.4] on the dynamics perturbed by the presence of external electromagnetic fields.

Remark 4.13
All results of this subsection depend on Theorem 4.8 (iii), i.e., the rate of convergence, as $n \to \infty$, of the family $\{\tau^{(n,x)}\}_{n \in \mathbb{N}_0}$ of finite–volume groups introduced in the

preliminary discussions before Lemma 4.9. It is the only information on the Fermi system we needed here.

Remark 4.14
The Lieb–Robinson bound for multi–commutators given by Theorems 4.10–4.11 at $k = 1$ is not as good as the previous Lieb–Robinson bound of Theorem 4.8 (iv). Nevertheless, they are qualitatively equivalent in the following sense: For interactions with polynomial decay, the first bound also has polynomial decay, even if with lower degree than the second one. For interactions with exponential decay, both bounds are exponentially decaying, even if the first one has a worse prefactor and exponential rate than the second one.

4.5 Application to Perturbed Autonomous Dynamics

Let $\Psi \in \mathcal{W}$ and $\mathbf{V}$ be a potential. For any $l \in \mathbb{R}_0^+$, we consider a map $\eta \mapsto \mathbf{W}^{(l,\eta)}$ from $\mathbb{R}$ to the subspace of self–adjoint elements of $\mathcal{U}_{\Lambda_l}$. In the case that interests us, the following property holds:

$$\left\| \mathbf{W}^{(l,\eta)} \right\|_{\mathcal{U}} = \mathcal{O}(\eta \, |\Lambda_l|) . \tag{4.67}$$

More precisely, we consider elements $\mathbf{W}^{(l,\eta)}$ of the form

$$\mathbf{W}^{(l,\eta)} \doteq \sum_{x \in \Lambda_l} \sum_{z \in \mathfrak{L}, |z| \leq 1} \mathbf{w}_{x,x+z}(\eta) a_x^* a_{x+z} , \qquad l \in \mathbb{R}_0^+ , \tag{4.68}$$

where $\{\mathbf{w}_{x,y}\}_{x,y \in \mathfrak{L}}$ are complex–valued functions of $\eta \in \mathbb{R}$ with

$$\overline{\mathbf{w}_{x,y}} = \mathbf{w}_{y,x} \qquad \text{and} \qquad \mathbf{w}_{x,y}(0) = 0 \tag{4.69}$$

for all $x, y \in \mathfrak{L}$.
Equation (4.68) has the form

$$\mathbf{W}^{(l,\eta)} = \sum_{x \in \Lambda_l} W_x(\eta) \tag{4.70}$$

where, for some fixed radius $R \in \mathbb{R}^+$ and any $x \in \mathfrak{L}$, $W_x(\eta)$ is a self–adjoint even element of $\mathcal{U}_{x+\Lambda_R}$ that depends on the real parameter η. All results below in this subsection hold for the more general case (4.70) as well, with obvious modifications. Indeed, we could even consider more general perturbations with $R = \infty$, see proofs of Inequality (5.33) and Theorem 5.6.

We refrain from treating cases more general than (4.68) to keep technical aspects as simple as possible. Observe that perturbations due to the presence of external electromagnetic fields are included in the class of perturbations defined by (4.68).

In fact, as discussed in the introduction, our final aim is the microscopic quantum theory of electrical conduction [BP4, BP5, BP6]. Indeed, at fixed $l \in \mathbb{R}_0^+$, $\mathbf{W}^{(l,\eta)}$ defined by (4.68) is related to perturbations of dynamics caused by constant external electromagnetic fields that vanish outside the box Λ_l.

We assume that $\{\mathbf{w}_{x,y}\}_{x,y\in\mathfrak{L}}$ are uniformly bounded and Lipschitz continuous: There is a constant $K_1 \in \mathbb{R}^+$ such that, for all $\eta, \eta_0 \in \mathbb{R}$,

$$\sup_{x,y\in\mathfrak{L}} \left| \mathbf{w}_{x,y}(\eta) - \mathbf{w}_{x,y}(\eta_0) \right| \leq K_1 \left| \eta - \eta_0 \right| \quad \text{and} \quad \sup_{x,y\in\mathfrak{L}} \sup_{\eta\in\mathbb{R}} \left| \mathbf{w}_{x,y}(\eta) \right| \leq K_1 .$$

$$(4.71)$$

These two uniformity conditions could hold for parameters η, η_0 on compact sets only, but we refrain again from considering this more general case, for simplicity.

The perturbed dynamics is defined via the symmetric derivation

$$\delta^{(l,\eta)} \doteq \delta + i \left[\mathbf{W}^{(l,\eta)}, \cdot \right] , \qquad l \in \mathbb{R}_0^+, \ \eta \in \mathbb{R} .$$

$$(4.72)$$

Recall that δ is the symmetric derivation of Theorem 4.8 which generates the C_0–group $\{\tau_t\}_{t\in\mathbb{R}}$ on $\mathcal{U}$. The second term in the r.h.s. of (4.72) is a bounded perturbation of δ. Hence, $\delta^{(l,\eta)}$ generates a C_0–group $\{\tilde{\tau}_t^{(l,\eta)}\}_{t\in\mathbb{R}}$ on $\mathcal{U}$, see [EN, Chap. III, Sect. 1.3]. By Lemma 4.6, the (generally unbounded) closed operator $\delta^{(l,\eta)}$ is a conservative symmetric derivation and $\tilde{\tau}_t^{(l,\eta)}$ is a $*$–automorphism of $\mathcal{U}$ for all $t \in \mathbb{R}$.

Let Φ be any interaction with energy observables

$$U_{\Lambda_L}^\Phi \doteq \sum_{\Lambda\subseteq\Lambda_L} \Phi_\Lambda , \qquad L \in \mathbb{R}_0^+ .$$

$$(4.73)$$

The main aim of this subsection is to study the energy increment

$$\mathbf{T}_{t,s}^{(l,\eta,L)} \doteq \tilde{\tau}_{t-s}^{(l,\eta)}(U_{\Lambda_L}^\Phi) - \tau_{t-s}(U_{\Lambda_L}^\Phi) , \qquad l, L \in \mathbb{R}_0^+, \ s,t,\eta \in \mathbb{R} ,$$

$$(4.74)$$

in the limit $L \to \infty$ to obtain similar results as [BPH1, Theorem 3.4]. This can be done by using the (partial) Dyson–Phillips series:

$$\mathbf{T}_{t,s}^{(l,\eta,L)} - \mathbf{T}_{t,s}^{(l,\eta_0,L)} \tag{4.75}$$

$$= \sum_{k=1}^m i^k \int_s^t ds_1 \cdots \int_s^{s_{k-1}} ds_k \left[\mathbf{X}_{s_k,s}^{(l,\eta_0,\eta)}, \ldots, \mathbf{X}_{s_1,s}^{(l,\eta_0,\eta)}, \tilde{\tau}_{t-s}^{(l,\eta_0)}(U_{\Lambda_L}^\Phi) \right]^{(k+1)}$$

$$+ i^{m+1} \int_s^t ds_1 \cdots \int_s^{s_m} ds_{m+1}$$

$$\tilde{\tau}_{s_{m+1}-s}^{(l,\eta)} \left(\left[\mathbf{W}^{(l,\eta)} - \mathbf{W}^{(l,\eta_0)}, \mathbf{X}_{s_m,s_{m+1}}^{(l,\eta_0,\eta)}, \ldots, \mathbf{X}_{s_1,s_{m+1}}^{(l,\eta_0,\eta)}, \tilde{\tau}_{t-s_{m+1}}^{(l,\eta_0)}(U_{\Lambda_L}^\Phi) \right]^{(m+2)} \right)$$

for any $m \in \mathbb{N}$, where

$$\mathbf{X}_{t,s}^{(l,\eta_0,\eta)} \doteq \tilde{\tau}_{t-s}^{(l,\eta_0)}(\mathbf{W}^{(l,\eta)} - \mathbf{W}^{(l,\eta_0)}) \,, \qquad l \in \mathbb{R}_0^+, \ s,t,\eta_0,\eta \in \mathbb{R} \,. \tag{4.76}$$

By (4.69), note that $\mathbf{T}_{t,s}^{(l,0,L)} = 0$.

By (4.67), naive bounds on the r.h.s. of (4.75) predict that

$$\left[\mathbf{X}_{s_k,s}^{(l,\eta_0,\eta)}, \ldots, \mathbf{X}_{s_1,s}^{(l,\eta_0,\eta)}, \tilde{\tau}_{t-s}^{(l,\eta_0)}(U_{\Lambda_L}^{\Phi})\right]^{(k+1)} = \mathcal{O}(|\Lambda_l|^k |\Lambda_L|) \,.$$

To obtain more accurate estimates, we use the tree–decay bounds on multi–commutators of Corollary 4.12.

To this end, for any $x \in \mathfrak{L}$ and $m \in \mathbb{N}$, we define

$$\mathcal{D}(x,m) \doteq \left\{\Lambda \in \mathcal{P}_f(\mathfrak{L}) : x \in \Lambda, \ \Lambda \subseteq \Lambda_m + x, \ \Lambda \nsubseteq \Lambda_{m-1} + x\right\} \subset 2^{\mathfrak{L}} \,. \tag{4.77}$$

All elements of $\mathcal{D}(x,m)$ are finite subsets of the lattice $\mathfrak{L}$ that contain at least two sites which are separated by a distance greater or equal than m. Using, for any $x \in \mathfrak{L}$ and $m = 0$, the convention

$$\mathcal{D}(x,0) \doteq \{\{x\}\} \,, \tag{4.78}$$

we obviously have that

$$\mathcal{P}_f(\mathfrak{L}) = \bigcup_{x \in \mathfrak{L}, \ m \in \mathbb{N}_0} \mathcal{D}(x,m) \,. \tag{4.79}$$

We now consider the following assumption on interactions Φ:

$$\sup_{x \in \mathfrak{L}} \sum_{m \in \mathbb{N}_0} \mathbf{v}_m \sum_{\Lambda \in \mathcal{D}(x,m)} \|\Phi_\Lambda\|_{\mathcal{U}} < \infty \tag{4.80}$$

for some (generally diverging) sequence $\{\mathbf{v}_m\}_{m \in \mathbb{N}_0} \subset \mathbb{R}_0^+$. For instance, if $\Phi \in \mathcal{W}$ and Condition (4.56) holds true, then one easily verifies (4.80) with $\mathbf{v}_m = (1+m)^\varsigma$. In the case (4.57) holds and $\Phi \in \mathcal{W}$, then (4.80) is also satisfied even with $\mathbf{v}_m = e^{m\varsigma}$.

We are now in position to state the first main result of this section, which is an extension of [BPH1, Theorem 3.4 (i)] to interacting fermions:

Theorem 4.15 (Taylor's theorem for increments)
Let $l, \mathrm{T} \in \mathbb{R}_0^+$, $s,t \in [-\mathrm{T}, \mathrm{T}]$, $\eta, \eta_0 \in \mathbb{R}$, $\Psi \in \mathcal{W}$, and $\mathbf{V}$ be any potential. Assume (4.56) with $\varsigma > d$, (4.69) and (4.71). Take an interaction Φ satisfying (4.80) with $\mathbf{v}_m = (1+m)^\varsigma$. Then:
(i) The map $\eta \mapsto \mathbf{T}_{t,s}^{(l,\eta,L)}$ converges uniformly on $\mathbb{R}$, as $L \to \infty$, to a continuous function $\mathbf{T}_{t,s}^{(l,\eta)}$ of η and

$$\mathbf{T}_{t,s}^{(l,\eta)} - \mathbf{T}_{t,s}^{(l,\eta_0)} = \sum_{\Lambda \in \mathcal{P}_f(\mathfrak{L})} i \int_s^t ds_1 \tilde{\tau}_{s_1-s}^{(l,\eta)}\left(\left[\mathbf{W}^{(l,\eta)} - \mathbf{W}^{(l,\eta_0)}, \tilde{\tau}_{t-s_1}^{(l,\eta_0)}(\Phi_\Lambda)\right]\right) \,.$$

(ii) *For any $m \in \mathbb{N}$ satisfying $d(m+1) < \varsigma$,*

$$\mathbf{T}_{t,s}^{(l,\eta)} - \mathbf{T}_{t,s}^{(l,\eta_0)} = \tag{4.81}$$

$$\sum_{k=1}^{m} \sum_{\Lambda \in \mathcal{P}_f(\mathfrak{L})} i^k \int_s^t \mathrm{d}s_1 \cdots \int_s^{s_{k-1}} \mathrm{d}s_k \left[\mathbf{X}_{s_k,s}^{(l,\eta_0,\eta)}, \ldots, \mathbf{X}_{s_1,s}^{(l,\eta_0,\eta)}, \tilde{\tau}_{t-s}^{(l,\eta_0)}(\Phi_\Lambda) \right]^{(k+1)}$$

$$+ \sum_{\Lambda \in \mathcal{P}_f(\mathfrak{L})} i^{m+1} \int_s^t \mathrm{d}s_1 \cdots \int_s^{s_m} \mathrm{d}s_{m+1}$$

$$\tilde{\tau}_{s_{m+1}-s}^{(l,\eta)} \left(\left[\mathbf{W}^{(l,\eta)} - \mathbf{W}^{(l,\eta_0)}, \mathbf{X}_{s_m,s_{m+1}}^{(l,\eta_0,\eta)}, \ldots, \mathbf{X}_{s_1,s_{m+1}}^{(l,\eta_0,\eta)}, \tilde{\tau}_{t-s_{m+1}}^{(l,\eta_0)}(\Phi_\Lambda) \right]^{(m+2)} \right).$$

(iii) *All the above series in Λ absolutely converge: For any $m \in \mathbb{N}$ satisfying $d(m+1) < \varsigma$, $k \in \{1, \ldots, m\}$, and $\{s_j\}_{j=1}^{m+1} \subset [-T, T]$,*

$$\sum_{\Lambda \in \mathcal{P}_f(\mathfrak{L})} \left\| \left[\mathbf{X}_{s_k,s}^{(l,\eta_0,\eta)}, \ldots, \mathbf{X}_{s_1,s}^{(l,\eta_0,\eta)}, \tilde{\tau}_{t-s}^{(l,\eta_0)}(\Phi_\Lambda) \right]^{(k+1)} \right\|_{\mathcal{U}} \leq D \, |\Lambda_l| \, |\eta - \eta_0|^k$$

and

$$\sum_{\Lambda \in \mathcal{P}_f(\mathfrak{L})} \left\| \tilde{\tau}_{s_{m+1}-s}^{(l,\eta)} \left(\left[\mathbf{W}^{(l,\eta)} - \mathbf{W}^{(l,\eta_0)}, \mathbf{X}_{s_m,s_{m+1}}^{(l,\eta_0,\eta)}, \ldots, \mathbf{X}_{s_1,s_{m+1}}^{(l,\eta_0,\eta)}, \tilde{\tau}_{t-s_{m+1}}^{(l,\eta_0)}(\Phi_\Lambda) \right]^{(m+2)} \right) \right\|_{\mathcal{U}}$$

$$\leq D \, |\Lambda_l| \, |\eta - \eta_0|^{m+1},$$

for some constant $D \in \mathbb{R}^+$ depending only on $m, d, \mathrm{T}, \Psi, K_1, \Phi, \mathbf{F}$. The last assertion also holds for $m = 0$.

Proof We only prove (ii)–(iii), Assertion (i) being easier to prove by very similar arguments. For simplicity, we assume w.l.o.g. $\eta_0 = s = 0$ and $m \in \mathbb{N}$. Because of Eqs. (4.68), (4.75), (4.76) and (4.79), we first control the multi–commutator sum

$$F_{k,L} \doteq \sum_{x_0 \in \mathfrak{L} \backslash \Lambda_L} \sum_{m_0 \in \mathbb{N}_0} \sum_{\Lambda \in \mathcal{D}(x_0,m_0)} \sum_{x_1 \in \Lambda_l} \sum_{z_1 \in \mathfrak{L}, |z_1| \leq 1} \cdots \sum_{x_k \in \Lambda_l} \sum_{z_k \in \mathfrak{L}, |z_k| \leq 1} \left\| \xi_{x_1,z_1,\ldots,x_k,z_k} \left[\tau_{s_k}(a_{x_k}^* a_{x_k+z_k}), \ldots, \tau_{s_1}(a_{x_1}^* a_{x_1+z_1}), \tau_t(\Phi_\Lambda) \right]^{(k+1)} \right\|_{\mathcal{U}}$$

for any fixed $k \in \{1, \ldots, m\}$, $\mathrm{T} \in \mathbb{R}_0^+$, $\{s_j\}_{j=1}^k \subset [-\mathrm{T}, \mathrm{T}]$ and $L \in \mathbb{R}_0^+ \cup \{-1\}$, where we use the convention $\Lambda_{-1} \doteq \emptyset$ and

$$\xi_{x_1,z_1,\ldots,x_k,z_k} \doteq \prod_{j=1}^{k} \mathbf{w}_{x_j,x_j+z_j}(\eta). \tag{4.82}$$

By (4.69)–(4.71), there is a constant $D \in \mathbb{R}^+$ (depending on K_1) such that

$$\sup_{x_1,z_1,\dots,x_k,z_k\in\mathcal{L}}\ \sup_{\eta\in\mathbb{R}}\left|\xi_{x_1,z_1,\dots,x_k,z_k}\right|\le D\quad\text{and}\quad\sup_{x_1,z_1,\dots,x_k,z_k\in\mathcal{L}}\left|\xi_{x_1,z_1,\dots,x_k,z_k}\right|\le D|\eta|^k\ .$$

$$(4.83)$$

At fixed $k\in\{1,\dots,m\}$ observe further that the condition $\varsigma>dk$ yields

$$\max_{x\in\mathcal{L}}\sum_{y\in\mathcal{L}}\frac{1}{(1+|y-x|)^{\varsigma(\max\{\partial_T(j),\partial_T(l)\})^{-1}}}\le\sum_{y\in\mathcal{L}}\frac{1}{(1+|y|)^{\frac{\varsigma}{k}}}<\infty\qquad(4.84)$$

for any tree $T\in\mathcal{T}_{k+1}$ and all $j,l\in\{0,\dots,k\}$. Using (4.80) with $\mathbf{v}_m=(1+m)^\varsigma$, (4.83)–(4.84) and the equality

$$\left\|\left[\tau_{s_k}(a_{x_k}^*a_{x_k+z_k}),\dots,\tau_{s_1}(a_{x_1}^*a_{x_1+z_1}),\tau_t(\Phi_\Lambda)\right]^{(k+1)}\right\|_{\mathcal{U}}$$
$$=\left\|\left[\tau_{s_k-t}(a_{x_k}^*a_{x_k+z_k}),\dots,\tau_{s_1-t}(a_{x_1}^*a_{x_1+z_1}),\Phi_\Lambda\right]^{(k+1)}\right\|_{\mathcal{U}},\qquad(4.85)$$

we obtain from Corollary 4.12 that, for any $m\in\mathbb{N}$ and $k\in\{1,\dots,m\}$ with $\varsigma>dk$, $F_{k,-1}\le D|\Lambda_l||\eta|^k$ for some constant $D\in\mathbb{R}^+$ depending only on m,d,T,Ψ, $K_1,\Phi,\mathbf{F}$.

Hence, by Lebesgue's dominated convergence theorem, for any $k\in\mathbb{N}$ satisfying $\varsigma>dk$, there is $R\in\mathbb{R}^+$ such that $F_{k,L}<\varepsilon$ for any $L\ge R$. This ensures the convergence of the first k multi–commutators of (4.75) to the first k multi–commutators of (4.81) as well as the corresponding absolute summability. Cf. Assertions (ii)–(iii). The convergence is even uniform for $\eta\in\mathbb{R}$ because of the first assertion of (4.83).

Because $\tilde{\tau}_t^{(l,\eta)}$ is an isometry for any time $t\in\mathbb{R}$, the same arguments are used to control the multi–commutator

$$\tilde{\tau}_{s_{m+1}-s}^{(l,\eta)}\left(\left[\mathbf{W}^{(l,\eta)}-\mathbf{W}^{(l,\eta_0)},\mathbf{X}_{s_m,s_{m+1}}^{(l,\eta_0,\eta)},\dots,\mathbf{X}_{s_1,s_{m+1}}^{(l,\eta_0,\eta)},\tilde{\tau}_{t-s_{m+1}}^{(l,\eta_0)}(\Phi_\Lambda)\right]^{(m+2)}\right)\qquad(4.86)$$

in (4.75). By (4.71), notice additionally that there is a constant $D\in\mathbb{R}^+$ and a family $\{\Psi^{(l,\eta)}\}_{l\in\mathbb{R}_0^+,\eta\in\mathbb{R}}\subset\mathcal{W}$ such that

$$\sup_{\eta\in\mathbb{R}}\ \sup_{l\in\mathbb{R}_0^+}\left\|\Psi^{(l,\eta)}\right\|_{\mathcal{W}}\le D<\infty$$

and, for all $l\in\mathbb{R}_0^+$ and $\eta\in\mathbb{R}$, $\{\tilde{\tau}_t^{(l,\eta)}\}_{t\in\mathbb{R}}$ is the C_0–group of $*$–automorphisms on $\mathcal{U}$ associated with the interaction $\Psi^{(l,\eta)}$ and the potential $\mathbf{V}$. The norm $\|\cdot\|_{\mathcal{W}}$ in the last inequality, which defines the space $\mathcal{W}$ of interactions, is of course defined w.r.t. the same function $\mathbf{F}$ to which the conditions of the theorem are imposed. This property justifies the simplifying assumption $\eta_0=0$ at the beginning of the proof. This concludes the proof of Assertions (ii)–(iii).

Assertion (i) is proven in the same way and we omit the details. Note only that the convergence of $F_{1,L}$ as $L\to\infty$ is uniform for $\eta\in\mathbb{R}$ because of the first assertion of (4.83). The latter implies the continuity of the map $\eta\mapsto\mathbf{T}_{t,s}^{(l,\eta)}$ for $\eta\in\mathbb{R}$.   ∎

A direct consequence of Theorem 4.15 is that $\mathbf{T}_{t,s}^{(l,\eta)} = \mathcal{O}(|\Lambda_l|)$. Note furthermore that Theorem 4.15 also holds when the cubic box Λ_l is replaced by *any* finite subset $\Lambda \in \mathcal{P}_f(\mathfrak{L})$. The assumptions of this theorem are fulfilled for any interactions $\Psi, \Phi \in \mathcal{W}$ with the decay function (4.7), provided the parameter $\epsilon \in \mathbb{R}^+$ is sufficiently large. Theorem 4.15 is thus a *significant* extension of [BPH1, Theorem 3.4 (i)] in the sense that very general inter–particle interactions and the full range of parameters $\eta \in \mathbb{R}$ are now allowed.

In the case of exponentially decaying interactions we can bound the derivatives $|\Lambda_l|^{-1}\partial_\eta^m \mathbf{T}_{t,s}^{(l,\eta)}$ for all $m \in \mathbb{N}$, uniformly w.r.t. $l \in \mathbb{R}_0^+$. We thus extend [BPH1, Theorem 3.4 (ii)] for interactions Φ satisfying (4.80).

Under these conditions, we show below that the map $\eta \mapsto |\Lambda_l|^{-1}\mathbf{T}_{t,s}^{(l,\eta)}$ from $\mathbb{R}$ to $\mathcal{U}$ is bounded in the sense of Gevrey norms, uniformly w.r.t. $l \in \mathbb{R}_0^+$. Note that real analytic functions (cf. [BPH1, Theorem 3.4 (ii)]) are a special case of Gevrey functions.

Theorem 4.16 (Increments as Gevrey maps)
Let $l, \mathrm{T} \in \mathbb{R}_0^+$, $s, t \in [-\mathrm{T}, \mathrm{T}]$, $\Psi \in \mathcal{W}$, and $\mathbf{V}$ be any potential. Assume (4.57) and take an interaction Φ satisfying (4.80) with $\mathbf{v}_m = e^{ms}$. Assume further the real analyticity of the maps $\eta \mapsto \mathbf{w}_{x,y}(\eta)$, $x, y \in \mathfrak{L}$, from $\mathbb{R}$ to $\mathbb{C}$ as well as the existence of $r \in \mathbb{R}^+$ such that

$$K_2 \doteq \sup_{x,y\in\mathfrak{L}} \sup_{m\in\mathbb{N}} \sup_{\eta\in\mathbb{R}} \frac{r^m \partial_\eta^m \mathbf{w}_{x,y}(\eta)}{m!} < \infty \,. \tag{4.87}$$

(i) *Smoothness. As a function of $\eta \in \mathbb{R}$, $\mathbf{T}_{t,s}^{(l,\eta)} \in C^\infty(\mathbb{R}; \mathcal{U})$ and for any $m \in \mathbb{N}$,*

$$\partial_\eta^m \mathbf{T}_{t,s}^{(l,\eta)} = \sum_{k=1}^m \sum_{\Lambda\in\mathcal{P}_f(\mathfrak{L})} i^k \int_s^t ds_1 \cdots \int_s^{s_{k-1}} ds_k$$

$$\partial_\varepsilon^m \left[\mathbf{X}_{s_k,s}^{(l,\eta,\eta+\varepsilon)}, \ldots, \mathbf{X}_{s_1,s}^{(l,\eta,\eta+\varepsilon)}, \tilde{\tau}_{t,s}^{(l,\eta)}(\Phi_\Lambda) \right]^{(k+1)} \Bigg|_{\varepsilon=0} .$$

The above series in Λ are absolutely convergent.
(ii) *Uniform boundedness of the Gevrey norm of density of increments. There exist $\tilde{r} \equiv \tilde{r}_{d,\mathrm{T},\Psi,K_2,\mathbf{F}} \in \mathbb{R}^+$ and $D \equiv D_{\mathrm{T},\Psi,K_2,\Phi} \in \mathbb{R}^+$ such that, for all $l \in \mathbb{R}_0^+$, $\eta \in \mathbb{R}$ and $s, t \in [-\mathrm{T}, \mathrm{T}]$,*

$$\sum_{m\in\mathbb{N}} \frac{\tilde{r}^m}{(m!)^d} \sup_{l\in\mathbb{R}_0^+} \left\| |\Lambda_l|^{-1} \partial_\eta^m \mathbf{T}_{t,s}^{(l,\eta)} \right\|_{\mathcal{U}} \leq D \,.$$

Before giving the proof, note first that the assumptions of Theorem 4.16 are satisfied for any interactions $\Psi, \Phi \in \mathcal{W}$ with the decay function (4.58). Moreover, under conditions of Theorem 4.16, the family $\{|\Lambda_l|^{-1}\mathbf{T}_{t,s}^{(l,\eta)}\}_{l\in\mathbb{R}_0^+}$ of functions of the variable η at dimension $d = 1$ is uniformly bounded w.r.t. analytic norms. In particular, for $d = 1$ and any state $\varrho \in \mathcal{U}^*$, the limit of the increment density $|\Lambda_l|^{-1}\varrho(\mathbf{T}_{t,s}^{(l,\eta)})$, as

$l \to \infty$ (possibly along subsequences), is either identically vanishing for all $\eta \in \mathbb{R}$, or is different from zero for η outside a discrete subset of $\mathbb{R}$. Note that, by contrast, general non–vanishing Gevrey functions can have arbitrarily small support. We discuss this with more details at the end of Sect. 5.3.

We now conclude this subsection by proving Theorem 4.16. To this end, we need the following estimate:

Proposition 4.17
There is a constant $D \in \mathbb{R}^+$ such that, for all $k \in \mathbb{N}$,

$$\sum_{T \in \mathcal{T}_{k+1}} \max_{j \in \{0,\dots,k\}} \max_{x_j \in \mathcal{L}} \sum_{x_0,\dots,\widehat{x}_j,\dots,x_k \in \mathcal{L}} \prod_{\{p,l\} \in T} e^{-\frac{s|x_p - x_l|}{\sqrt{d}\max\{\partial_T(p),\partial_T(l)\}}} \le D^k (k!)^d .$$

The proof of this upper bound uses the fact that trees with vertices of large degree are *"rare"* in a way that summing up the numbers $(\partial_T!)^\alpha$ for $T \in \mathcal{T}_{k+1}$ and any $\alpha \in \mathbb{R}^+$ gives factors behaving, at worst, like $D^k (k!)^\alpha$. The arguments are standard results of finite mathematics. We prove them below for completeness, in two simple lemmata.

Let $k \in \mathbb{N}$. For any fixed sequence $\partial = (\partial(0), \dots, \partial(k)) \in \mathbb{N}^{k+1}$ define the set $\mathcal{T}_{k+1}(\partial) \subset \mathcal{T}_{k+1}$ by

$$\mathcal{T}_{k+1}(\partial) \doteq \{T \in \mathcal{T}_{k+1} : \partial_T \equiv (\partial_T(0), \dots, \partial_T(k)) = \partial\} .$$

In other words, $\mathcal{T}_{k+1}(\partial)$ is the set of all trees of $\mathcal{T}_{k+1}$ with vertices having their degree fixed by the sequence ∂. The cardinality of this set is bounded as follows:

Lemma 4.18 (Number of trees with vertices of fixed degrees)
For all $k \in \mathbb{N}$ and $\partial \in \mathbb{N}^{k+1}$,

$$|\mathcal{T}_{k+1}(\partial)| \le \frac{(k-1)!}{(\partial(0) - 1)! \cdots (\partial(k) - 1)!} .$$

Proof The bound can be proven, for instance, by using so–called "Prüfer codes". We give here a proof based on a simplified version of such codes, well adapted to the particular sets of trees $\mathcal{T}_{k+1}$. At fixed $k \in \mathbb{N}$, define the map $\mathfrak{C} : \mathcal{T}_{k+1} \to \{0, \dots, k-1\}^{k-1}$ by

$$\mathfrak{C}(T) \doteq (\mathrm{P}_T(2), \dots, \mathrm{P}_T(k)) .$$

See (4.45)–(4.47). This map is clearly injective and if $j \in \{0, \dots, k\}$ is a vertex of degree $\partial_T(j)$, then it appears exactly $(\partial_T(j) - 1)$ times in the sequence $\mathfrak{C}(T)$. Note that $\partial_T(k) = 1$ for all $T \in \mathcal{T}_{k+1}$. To finish the proof, fix $\partial = (\partial(0), \dots, \partial(k)) \in \mathbb{N}^{k+1}$ and observe that if $\partial(0) + \cdots + \partial(k) = 2k$ then there are exactly

$$\frac{(k-1)!}{(\partial(0) - 1)! \cdots (\partial(k) - 1)!}$$

sequences in $\{0, \ldots, k-1\}^{k-1}$ with $j \in \{0, \ldots, k\}$ appearing exactly $(\eth(j)-1)$ times in such sequences. If $\eth(0) + \cdots + \eth(k) \neq 2k$ then such a sequence does not exist.   ∎

Lemma 4.19

For all $k \in \mathbb{N}$,

$$\sum_{\eth(0),\ldots,\eth(k)\in\mathbb{N}} \mathbb{1}[\eth(0) + \cdots + \eth(k) = 2k] \leq 4^k .$$

Proof For $k \in \mathbb{N}$, the coefficient c_{2k} of the analytic function

$$z \mapsto \frac{z^{k+1}}{(1-z)^{k+1}} = \sum_{m=1}^{\infty} c_m z^m$$

on the complex disc $\{z \in \mathbb{C} : |z| < 1\}$ is exactly the finite sum

$$\sum_{\eth(0),\ldots,\eth(k)\in\mathbb{N}} \mathbb{1}[\eth(0) + \cdots + \eth(k) = 2k] .$$

In particular,

$$\sum_{\eth(0),\ldots,\eth(k)\in\mathbb{N}} \mathbb{1}[\eth(0) + \cdots + \eth(k) = 2k] = \frac{1}{2\pi i} \oint_{|z|=1/2} \frac{1}{z^k(1-z)^{k+1}} dz ,$$

which combined with the inequality

$$\left| \frac{1}{2\pi i} \oint_{|z|=1/2} \frac{1}{z^k(1-z)^{k+1}} dz \right| \leq 4^k$$

yields the assertion.   ∎

By using the two above lemmata, we now prove Proposition 4.17:

Proof Fix $\alpha \in \mathbb{R}^+$ and note first that, for all $d \in \mathbb{N}$,

$$\lim_{g \to \infty} \frac{1}{g^d} \sum_{x \in \mathfrak{L}} e^{-\frac{\alpha|x|}{g\sqrt{d}}} = \int_{\mathbb{R}^d} e^{-\frac{\alpha|x|}{\sqrt{d}}} d^d x < \infty .$$

Hence, for $d \in \mathbb{N}$, there is a constant $S_d \in \mathbb{R}^+$ such that

$$\sum_{x \in \mathfrak{L}} e^{-\frac{\alpha|x|}{g\sqrt{d}}} \leq S_d g^d , \qquad g \in \mathbb{N} .$$

From this estimate and by using the Stirling–type bounds [Ro]

$$g^g e^{-g} e^{\frac{1}{12g+1}} \sqrt{2\pi g} \leq g! \leq g^g e^{-g} e^{\frac{1}{12g}} \sqrt{2\pi g} \,, \qquad g \in \mathbb{N} \,, \tag{4.88}$$

we obtain

$$\max_{j \in \{0,\dots,k\}} \max_{x_j \in \mathfrak{L}} \sum_{x_0,\dots,\not{x}_j,\dots,x_k \in \mathfrak{L}} \prod_{\{p,l\} \in T} \exp\left(-\frac{\varsigma \, |x_p - x_l|}{\sqrt{d} \max\{\eth_T(p), \eth_T(l)\}} \right)$$

$$\leq S_d^k \prod_{j=0}^{k} \eth_T(j)^{\eth_T(j)d} \leq S_d^k e^{\eth_T(j)d} (\eth_T!)^d \tag{4.89}$$

for all $d, k \in \mathbb{N}$ and $T \in \mathcal{T}_{k+1}$. We infer from (4.60) that

$$\sum_{T \in \mathcal{T}_{k+1}} (\eth_T!)^d \leq (k!)^{d-1} \sum_{T \in \mathcal{T}_{k+1}} (\eth_T!) \,. \tag{4.90}$$

We use now Lemma 4.18 to get

$$\sum_{T \in \mathcal{T}_{k+1}} (\eth_T!) = \sum_{\eth(0),\dots,\eth(k) \in \mathbb{N}} \mathbb{1}[\eth(0) + \cdots + \eth(k) = 2k] \sum_{T \in \mathcal{T}_{k+1}((\eth(0),\dots,\eth(k)))} (\eth_T!)$$

$$\leq k! \sum_{\eth(0),\dots,\eth(k) \in \mathbb{N}} \mathbb{1}[\eth(0) + \cdots + \eth(k) = 2k] \, \eth(0) \cdots \eth(k)$$

$$\leq k! \sum_{\eth(0),\dots,\eth(k) \in \mathbb{N}} \mathbb{1}[\eth(0) + \cdots + \eth(k) = 2k] \, e^{\eth(0)} \cdots e^{\eth(k)} \,.$$

We invoke (4.59) and Lemma 4.19 to arrive at

$$\sum_{T \in \mathcal{T}_{k+1}} (\eth_T!) \leq (k!) e^{2k} \sum_{\eth(0),\dots,\eth(k) \in \mathbb{N}} \mathbb{1}[\eth(0) + \cdots + \eth(k) = 2k] \leq (k!)(4e^2)^k \,.$$

$$\tag{4.91}$$

Proposition 4.17 is then a consequence of (4.89), (4.90) and (4.91).   ∎

We are now in position to prove Theorem 4.16:

Proof (i) Observe that

$$\partial_\eta^m \mathbf{T}_{t,s}^{(l,\eta,L)} = \partial_\varepsilon^m \left(\mathbf{T}_{t,s}^{(l,\eta+\varepsilon,L)} - \mathbf{T}_{t,s}^{(l,\eta,L)} \right)\Big|_{\varepsilon=0} \,. \tag{4.92}$$

The difference $\mathbf{T}_{t,s}^{(l,\eta+\varepsilon,L)} - \mathbf{T}_{t,s}^{(l,\eta,L)}$ is explicitly given by a Dyson–Phillips series involving multi–commutators (4.37)–(4.38): Use (4.75) to produce an infinite series. As the function $\eta \mapsto \mathbf{W}^{(l,\eta)}$ is, by assumption, real analytic, it follows that

$$\partial_\varepsilon^m \left(\mathbf{T}_{t,s}^{(l,\eta+\varepsilon,L)} - \mathbf{T}_{t,s}^{(l,\eta,L)} \right) \Big|_{\varepsilon=0} = \tag{4.93}$$

$$\sum_{k=1}^m i^k \int_s^t ds_1 \cdots \int_s^{s_{k-1}} ds_k \partial_\varepsilon^m \left[\mathbf{X}_{s_k,s}^{(l,\eta,\eta+\varepsilon)}, \ldots, \mathbf{X}_{s_1,s}^{(l,\eta,\eta+\varepsilon)}, \tilde{\tau}_{t,s}^{(l,\eta)}(U_{\Lambda_L}^\Phi) \right]^{(k+1)} \Big|_{\varepsilon=0}$$

for any $m \in \mathbb{N}$, $l \in \mathbb{R}_0^+$, and $s, t, \eta \in \mathbb{R}$. Set

$$\xi_{x_1,z_1,\ldots,x_k,z_k} \doteq \partial_\varepsilon^m \left\{ \prod_{j=1}^k \left(\mathbf{w}_{x_j,x_j+z_j}(\eta+\varepsilon) - \mathbf{w}_{x_j,x_j+z_j}(\eta) \right) \right\} \Big|_{\varepsilon=0}.$$

By (4.87), these coefficients are uniformly bounded w.r.t. $x_1, z_1, \ldots, x_k, z_k$ and η:

$$\sup_{x_1,z_1,\ldots,x_k,z_k\in\mathfrak{L}} \sup_{\eta\in\mathbb{R}} |\xi_{x_1,z_1,\ldots,x_k,z_k}| \le D^m m! \tag{4.94}$$

for some constant $D \in \mathbb{R}^+$ depending on K_2 but not on $m \ge k$. Bounding the above multi–commutators exactly as done for the proof of Theorem 4.15 and by taking the limit $L \to \infty$, we deduce from (4.92)–(4.93) that, for any $m \in \mathbb{N}$ and $s, t, \eta \in \mathbb{R}$,

$$\lim_{L\to\infty} \partial_\eta^m \mathbf{T}_{t,s}^{(l,\eta,L)} = \sum_{k=1}^m \sum_{\Lambda\in\mathcal{P}_f(\mathfrak{L})} i^k \int_s^t ds_1 \cdots \int_s^{s_{k-1}} ds_k \tag{4.95}$$

$$\partial_\varepsilon^m \left[\mathbf{X}_{s_k,s}^{(l,\eta,\eta+\varepsilon)}, \ldots, \mathbf{X}_{s_1,s}^{(l,\eta,\eta+\varepsilon)}, \tilde{\tau}_{t,s}^{(l,\eta)}(\Phi_\Lambda) \right]^{(k+1)} \Big|_{\varepsilon=0}.$$

This limit is uniform for $\eta \in \mathbb{R}$ because of (4.94). As in Theorem 4.15 (ii), the above series in Λ are absolutely convergent. Moreover, the uniform convergence of $\partial_\eta^m \mathbf{T}_{t,s}^{(l,\eta,L)}$, $m \in \mathbb{N}$, together with Theorem 4.15 (i) implies that the energy increment limit $\mathbf{T}_{t,s}^{(l,\eta)}$ is a smooth function of η with m–derivatives

$$\partial_\eta^m \mathbf{T}_{t,s}^{(l,\eta)} = \lim_{L\to\infty} \partial_\eta^m \mathbf{T}_{t,s}^{(l,\eta,L)}$$

for all $m \in \mathbb{N}$ and $s, t, \eta \in \mathbb{R}$. Because of (4.95), Assertion (i) thus follows.
(ii) is a direct consequence of (i), Corollary 4.12, and Proposition 4.17 together with (4.94) and

$$\int_s^t ds_1 \cdots \int_s^{s_{k-1}} ds_k \le \frac{(2T)^k}{k!}. \qquad \blacksquare$$

Chapter 5
Lieb–Robinson Bounds for Non-autonomous Dynamics

Like in Sect. 4, we only consider fermion systems, but all results can easily be extended to quantum spin systems (Sect. 3.6). For quantum spin systems, note that Lieb–Robinson bounds for non-autonomous dynamics have already been considered in [BMNS]. However, [BMNS] only proves Lieb–Robinson bounds for commutators, while the multi-commutator case was not considered, in contrast with results of this section. Observe also that some aspects of the non-autonomous case can be treated in a similar way to the autonomous case. However, several important arguments cannot be directly extended to the non-autonomous situation. Here, we only address in detail the technical issues which are specific to the non-autonomous problem. See for instance Corollary 5.2 (iii), Lemma 5.3, Theorems 5.5, and 5.7.

5.1 Existence of Non-autonomous Dynamics

We now consider time-dependent models. So, let $\Psi \doteq \{\Psi^{(t)}\}_{t \in \mathbb{R}}$ be a map from $\mathbb{R}$ to $\mathcal{W}$ such that

$$\|\Psi\|_\infty \doteq \sup_{t \in \mathbb{R}} \left\|\Psi^{(t)}\right\|_{\mathcal{W}} < \infty .$$

I.e., $\{\Psi^{(t)}\}_{t \in \mathbb{R}}$ is a *bounded* family in $\mathcal{W}$. We could easily extend the study of this section to families $\{\Psi^{(t)}\}_{t \in \mathbb{R}}$ which are only bounded for t on compacta. We refrain from considering this more general case, for simplicity. Take, furthermore, any collection $\{\mathbf{V}^{(t)}\}_{t \in \mathbb{R}}$ of potentials. Note that (4.10) is allowed for any $t \in \mathbb{R}$.

For all $x \in \mathfrak{L}$ and $\Lambda \in \mathcal{P}_f(\mathfrak{L})$, assume the continuity of the two maps $t \mapsto \Psi^{(t)}_\Lambda$, $t \mapsto \mathbf{V}^{(t)}_{\{x\}}$ from $\mathbb{R}$ to $\mathcal{U}$, i.e., $\Psi_\Lambda, \mathbf{V}_{\{x\}} \in C(\mathbb{R}; \mathcal{U})$. For any $L \in \mathbb{R}^+_0$, this yields the existence, uniqueness and an explicit expression, as a Dyson–Phillips series (cf. (2.11)), of the solution $\{\tau^{(L)}_{t,s}\}_{s,t \in \mathbb{R}}$ of the (finite-volume) non-autonomous evolutions equations

© The Author(s) 2017
J.-B. Bru and W. de Siqueira Pedra, *Lieb-Robinson Bounds for Multi-commutators and Applications to Response Theory*, SpringerBriefs in Mathematical Physics, DOI 10.1007/978-3-319-45784-0_5

$$\forall s, t \in \mathbb{R} : \qquad \partial_s \tau_{t,s}^{(L)} = -\delta_s^{(L)} \circ \tau_{t,s}^{(L)} , \qquad \tau_{t,t}^{(L)} = \mathbf{1}_{\mathcal{U}} , \tag{5.1}$$

and

$$\forall s, t \in \mathbb{R} : \qquad \partial_t \tau_{t,s}^{(L)} = \tau_{t,s}^{(L)} \circ \delta_t^{(L)} , \qquad \tau_{s,s}^{(L)} = \mathbf{1}_{\mathcal{U}} . \tag{5.2}$$

Here, for any $t \in \mathbb{R}$ and $L \in \mathbb{R}_0^+$, the bounded linear operator $\delta_t^{(L)}$ is defined on $\mathcal{U}$ by

$$\delta_t^{(L)}(B) \doteq i \sum_{\Lambda \subseteq \Lambda_L} \left[\Psi_\Lambda^{(t)}, B \right] + i \sum_{x \in \Lambda_L} \left[\mathbf{V}_{\{x\}}^{(t)}, B \right] , \qquad B \in \mathcal{U} .$$

Compare this definition with (4.3). As explained in Sect. 2.4 (see in particular Eqs. (2.15)–(2.16)), recall that the natural non-autonomous evolution equation in Quantum Mechanics is (5.2), but, by boundedness of $\delta_t^{(L)}$ for all times, (5.1) and (5.2) are both satisfied.

Similar to the autonomous case, for any $L \in \mathbb{R}_0^+$, $\{\tau_{t,s}^{(L)}\}_{s,t \in \mathbb{R}}$ is a continuous two-parameter family of bounded operators that satisfies the (reverse) cocycle property

$$\forall s, r, t \in \mathbb{R} : \qquad \tau_{t,s}^{(L)} = \tau_{r,s}^{(L)} \tau_{t,r}^{(L)} . \tag{5.3}$$

Its time-dependent generator $\delta_t^{(L)}$ is clearly a symmetric derivation and $\tau_{t,s}^{(L)}$ is thus a ∗-automorphism on $\mathcal{U}$ for all $L \in \mathbb{R}_0^+$ and $s, t \in \mathbb{R}$. Moreover, similar to the autonomous case (cf. Theorem 4.3 and Lemma 4.4), for all $L \in \mathbb{R}_0^+$ and $s, t \in \mathbb{R}$, $\tau_{t,s}^{(L)}$ satisfies Lieb–Robinson bounds and thus converges in the strong sense on $\mathcal{U}_0$, as $L \to \infty$:

Theorem 5.1 (Properties of non-autonomous finite-volume dynamics)
Let $\Psi \doteq \{\Psi^{(t)}\}_{t \in \mathbb{R}}$ be a bounded family on $\mathcal{W}$ (i.e., $\|\Psi\|_\infty < \infty$) and $\{\mathbf{V}^{(t)}\}_{t \in \mathbb{R}}$ a collection of potentials. For any $x \in \mathfrak{L}$ and $\Lambda \in \mathcal{P}_f(\mathfrak{L})$, assume $\Psi_\Lambda, \mathbf{V}_{\{x\}} \in C(\mathbb{R}; \mathcal{U})$. Fix $s, t \in \mathbb{R}$.
(i) Lieb–Robinson bounds. For any $L \in \mathbb{R}_0^+$, $B_1 \in \mathcal{U}^+ \cap \mathcal{U}_{\Lambda^{(1)}}$, and $B_2 \in \mathcal{U}_{\Lambda^{(2)}}$ with $\Lambda^{(1)}, \Lambda^{(2)} \subsetneq \Lambda_L$ and $\Lambda^{(1)} \cap \Lambda^{(2)} = \emptyset$,

$$\left\| \left[\tau_{t,s}^{(L)}(B_1), B_2 \right] \right\|_{\mathcal{U}}$$
$$\leq 2\mathbf{D}^{-1} \|B_1\|_{\mathcal{U}} \|B_2\|_{\mathcal{U}} \left(e^{2\mathbf{D}|t-s| \|\Psi\|_\infty} - 1 \right) \sum_{x \in \partial_\Psi \Lambda^{(1)}} \sum_{y \in \Lambda^{(2)}} \mathbf{F}(|x - y|) .$$

(ii) Convergence of the finite-volume dynamics. For any $\Lambda \in \mathcal{P}_f(\mathfrak{L})$, $B \in \mathcal{U}_\Lambda$, and $L_1, L_2 \in \mathbb{R}_0^+$ with $\Lambda \subset \Lambda_{L_1} \subsetneq \Lambda_{L_2}$,

$$\left\| \tau_{t,s}^{(L_2)}(B) - \tau_{t,s}^{(L_1)}(B) \right\|_{\mathcal{U}}$$
$$\leq 2 \|B\|_{\mathcal{U}} \|\Psi\|_\infty |t - s| e^{4\mathbf{D}|t-s| \|\Psi\|_\infty} \sum_{y \in \Lambda_{L_2} \setminus \Lambda_{L_1}} \sum_{x \in \Lambda} \mathbf{F}(|x - y|) .$$

Proof (i) The arguments are a straightforward extension of those proving Theorem 4.3 to non-autonomous dynamics: Fix $L \in \mathbb{R}_0^+$, $B_1 \in \mathcal{U}^+ \cap \mathcal{U}_{\Lambda^{(1)}}$ and $B_2 \in \mathcal{U}_{\Lambda^{(2)}}$ with disjoint sets $\Lambda^{(1)}, \Lambda^{(2)} \subsetneq \Lambda_L$. Similar to (4.13) and (4.14), we infer from (5.1) and (5.2) that the derivative w.r.t. to t of the function

$$f(s,t) \doteq \left[\tau_{t,s}^{(L)} \circ \tau_{s,t}^{(\Lambda^{(1)})} (B_1), B_2 \right], \qquad s,t \in \mathbb{R},$$

equals

$$\partial_t f(s,t) = i \sum_{Z \in \mathcal{S}_{\Lambda_L}(\Lambda^{(1)})} \left[\tau_{t,s}^{(L)}(\Psi_Z^{(t)}), f(s,t) \right] \tag{5.4}$$

$$- i \sum_{Z \in \mathcal{S}_{\Lambda_L}(\Lambda^{(1)})} \left[\tau_{t,s}^{(L)} \circ \tau_{s,t}^{(\Lambda^{(1)})} (B_1), \left[\tau_{t,s}^{(L)}(\Psi_Z^{(t)}), B_2 \right] \right].$$

Exactly like (4.15), it follows that

$$\| f(s,t) \|_{\mathcal{U}} \leq \| f(s,s) \|_{\mathcal{U}} + 2 \| B_1 \|_{\mathcal{U}} \sum_{Z \in \mathcal{S}_{\Lambda_L}(\Lambda^{(1)})} \int_{\min\{s,t\}}^{\max\{s,t\}} \left\| \left[\tau_{\alpha,s}^{(L)}(\Psi_Z^{(\alpha)}), B_2 \right] \right\|_{\mathcal{U}} d\alpha$$

for any $s, t \in \mathbb{R}$. Therefore, by using estimates that are similar to (4.16)–(4.18), we deduce Assertion (i).

(ii) The arguments are extensions to the non-autonomous case of those proving Lemma 4.4: Since $\Psi_\Lambda, \mathbf{V}_{\{x\}} \in C(\mathbb{R}; \mathcal{U})$ for any $x \in \mathcal{L}$ and $\Lambda \in \mathcal{P}_f(\mathcal{L})$, the time-dependent energy observables

$$H_L^{(t)} \doteq \sum_{\Lambda \subseteq \Lambda_L} \Psi_\Lambda^{(t)} + \sum_{x \in \Lambda_L} \mathbf{V}_{\{x\}}^{(t)}, \qquad L \in \mathbb{R}_0^+, \ t \in \mathbb{R},$$

and potentials

$$\mathbf{V}_Z^{(t)} \doteq \sum_{x \in Z} \mathbf{V}_{\{x\}}^{(t)} \in \mathcal{U}^+ \cap \mathcal{U}_Z, \qquad Z \in \mathcal{P}_f(\mathcal{L}), \ t \in \mathbb{R},$$

generate two solutions $\{\mathcal{V}_{s,t}(H_L)\}_{s,t \in \mathbb{R}}$ and $\{\mathcal{V}_{s,t}(\mathbf{V}_Z)\}_{s,t \in \mathbb{R}}$, respectively, of the non-autonomous evolution equations

$$\partial_t \left(\mathcal{V}_{s,t}(X) \right) = i \mathcal{V}_{s,t}(X) X^{(t)} \quad \text{and} \quad \partial_s \left(\mathcal{V}_{s,t}(X) \right) = -i X^{(s)} \mathcal{V}_{s,t}(X) \tag{5.5}$$

with $X^{(t)} = H_L^{(t)}$ or $\mathbf{V}_Z^{(t)}$. These evolution families satisfy $\mathcal{V}_{t,t}(X) = \mathbf{1}_{\mathcal{U}}$ for $t \in \mathbb{R}$ as well as the (usual) cocycle (Chapman–Kolmogorov) property

$$\forall t, r, s \in \mathbb{R}: \qquad \mathcal{V}_{s,t}(X) = \mathcal{V}_{s,r}(X) \mathcal{V}_{r,t}(X). \tag{5.6}$$

For any $L \in \mathbb{R}_0^+$ and $s, t, \alpha \in \mathbb{R}$, we then replace (4.19) in the proof of Lemma 4.4 with

$$\mathbf{U}_L(t, \alpha) \doteq \mathcal{V}_{s,t}(\mathbf{V}_{\Lambda_L})\mathcal{V}_{t,\alpha}(H_L)\mathcal{V}_{\alpha,s}(\mathbf{V}_{\Lambda_L}) . \tag{5.7}$$

By (5.6), $\mathbf{U}_L(t, t) = \mathbf{1}_{\mathcal{U}}$ for all $t \in \mathbb{R}$ while

$$\partial_t \mathbf{U}_L(t, \alpha) = -iG_L(t)\mathbf{U}_L(t, \alpha) \quad \text{and} \quad \partial_\alpha \mathbf{U}_L(t, \alpha) = i\mathbf{U}_L(t, \alpha) G_L(\alpha) \tag{5.8}$$

with

$$G_L(t) \doteq \sum_{\mathcal{Z} \subseteq \Lambda_L} \mathcal{V}_{s,t}(\mathbf{V}_{\Lambda_L}) \, \Psi_{\mathcal{Z}} \, \mathcal{V}_{t,s}(\mathbf{V}_{\Lambda_L}) . \tag{5.9}$$

Using the notation

$$\tilde{\tau}_{t,s}^{(L)}(B) \doteq \mathbf{U}_L(s, t) \, B \mathbf{U}_L(t, s) , \qquad B \in \mathcal{U}_\Lambda , \tag{5.10}$$

for any $s, t \in \mathbb{R}$ and $L \in \mathbb{R}_0^+$ such that $\Lambda \subset \Lambda_L$, observe that

$$\tau_{t,s}^{(L)}(B) = \mathcal{V}_{s,t}(H_L) B \mathcal{V}_{t,s}(H_L) = \tilde{\tau}_{t,s}^{(L)}\left(\mathcal{V}_{s,t}(\mathbf{V}_\Lambda) B \mathcal{V}_{t,s}(\mathbf{V}_\Lambda)\right) . \tag{5.11}$$

Note that, for any $s, t \in \mathbb{R}$, $\Lambda, \mathcal{Z} \in \mathcal{P}_f(\mathfrak{L})$ and $B \in \mathcal{U}_\Lambda$,

$$\mathcal{V}_{s,t}(\mathbf{V}_{\mathcal{Z}}) B \mathcal{V}_{t,s}(\mathbf{V}_{\mathcal{Z}}) \in \mathcal{U}_\Lambda \quad \text{and} \quad \left\| \mathcal{V}_{s,t}(\mathbf{V}_{\mathcal{Z}}) B \mathcal{V}_{t,s}(\mathbf{V}_{\mathcal{Z}}) \right\|_{\mathcal{U}} = \|B\|_{\mathcal{U}} . \tag{5.12}$$

Hence, it suffices to study the net $\{\tilde{\tau}_{t,s}^{(L)}(B)\}_{L \in \mathbb{R}_0^+}$ with $B \in \mathcal{U}_\Lambda$. Up to straightforward modifications taking into account the initial time $s \in \mathbb{R}$, the remaining part of the proof is now identical to the arguments starting from Eq. (4.20) in the proof of Lemma 4.4.   ∎

Corollary 5.2 (Infinite-volume dynamics)
Under the conditions of Theorem 5.1, finite-volume families $\{\tau_{t,s}^{(L)}\}_{s,t \in \mathbb{R}}$, $L \in \mathbb{R}_0^+$, converge strongly and uniformly for s, t on compact sets to a strongly continuous two-parameter family $\{\tau_{t,s}\}_{s,t \in \mathbb{R}}$ of $*$-automorphisms on $\mathcal{U}$ satisfying the following properties:
(i) *Reverse cocycle property.*

$$\forall s, r, t \in \mathbb{R} : \qquad \tau_{t,s} = \tau_{r,s}\tau_{t,r} .$$

(ii) *Lieb–Robinson bounds. For any $s, t \in \mathbb{R}$, $B_1 \in \mathcal{U}^+ \cap \mathcal{U}_{\Lambda^{(1)}}$, and $B_2 \in \mathcal{U}_{\Lambda^{(2)}}$ with disjoint sets $\Lambda^{(1)}, \Lambda^{(2)} \in \mathcal{P}_f(\mathfrak{L})$,*

$$\left\| [\tau_{t,s}(B_1), B_2] \right\|_{\mathcal{U}}$$
$$\leq 2\mathbf{D}^{-1} \|B_1\|_{\mathcal{U}} \|B_2\|_{\mathcal{U}} \left(e^{2\mathbf{D}|t-s|\|\Psi\|_\infty} - 1\right) \sum_{x \in \partial_\Psi \Lambda^{(1)}} \sum_{y \in \Lambda^{(2)}} \mathbf{F}(|x - y|) .$$

(iii) *Non-autonomous evolution equation. If* $\Psi \in C(\mathbb{R}; \mathcal{W})$ *then* $\{\tau_{t,s}\}_{s,t\in\mathbb{R}}$ *is the unique family of bounded operators on* $\mathcal{U}$ *satisfying, in the strong sense on the dense domain* $\mathcal{U}_0 \subset \mathcal{U}$,

$$\forall s, t \in \mathbb{R}: \quad \partial_t \tau_{t,s} = \tau_{t,s} \circ \delta_t, \quad \tau_{s,s} = \mathbf{1}_{\mathcal{U}}. \tag{5.13}$$

Here, δ_t, $t \in \mathbb{R}$, *are the conservative closed symmetric derivations, with common core* $\mathcal{U}_0$, *associated with the interactions* $\Psi^{(t)} \in \mathcal{W}$ *and the potentials* $\mathbf{V}^{(t)}$. *See Theorem 4.8.*

Proof The existence of a strongly continuous two-parameter family $\{\tau_{t,s}\}_{s,t\in\mathbb{R}}$ of *-automorphisms satisfying Lieb–Robinson bounds (ii) is a direct consequence of Theorem 5.1 together with the density of $\mathcal{U}_0 \subset \mathcal{U}$ and completeness of $\mathcal{U}$. This limiting family also satisfies the reverse cocycle property (i) because of (5.3).

(iii) For any $B \in \mathcal{U}_0 \subset \mathrm{Dom}(\delta_t)$, the map $t \mapsto \tau_{t,s} \circ \delta_t(B)$ from $\mathbb{R}$ to $\mathcal{U}$ is continuous. Indeed, for any $B \in \mathcal{U}_0$ and $\alpha, t \in \mathbb{R}$,

$$\left\| \tau_{\alpha,s} \circ \delta_\alpha(B) - \tau_{t,s} \circ \delta_t(B) \right\|_{\mathcal{U}} \leq \left\| (\tau_{\alpha,s} - \tau_{t,s}) \circ \delta_t(B) \right\|_{\mathcal{U}} + \left\| \delta_\alpha(B) - \delta_t(B) \right\|_{\mathcal{U}}.$$

By applying (4.28) to the interaction $\Psi^{(t)} - \Psi^{(\alpha)}$ and the potential $\mathbf{V}^{(t)} - \mathbf{V}^{(\alpha)}$ together with the strong continuity of $\{\tau_{t,s}\}_{s,t\in\mathbb{R}}$, one sees that, in the limit $\alpha \to t$, the r.h.s of the above inequality vanishes when $B \in \mathcal{U}_0$ and $\Psi \in C(\mathbb{R}; \mathcal{W})$. Now, because of (5.2), for any $L \in \mathbb{R}_0^+$, $B \in \mathcal{U}_0$, and $s, t \in \mathbb{R}$,

$$\left\| \tau_{t,s}(B) - B - \int_s^t \tau_{\alpha,s} \circ \delta_\alpha(B) \, d\alpha \right\|_{\mathcal{U}} \leq \left\| \tau_{t,s}(B) - \tau_{t,s}^{(L)}(B) \right\|_{\mathcal{U}} \tag{5.14}$$

$$+ \int_s^t \left\| (\tau_{\alpha,s}^{(L)} - \tau_{\alpha,s}) \circ \delta_\alpha(B) \right\|_{\mathcal{U}} d\alpha$$

$$+ \int_s^t \left\| \delta_\alpha^{(L)}(B) - \delta_\alpha(B) \right\|_{\mathcal{U}} d\alpha.$$

By using the strong convergence of $\tau_{t,s}^{(L)}$ towards $\tau_{t,s}$ as well as (4.28) and (4.80) together with Lebesgue's dominated convergence theorem, one checks that the r.h.s. of (5.14) vanishes when $B \in \mathcal{U}_0$ and $L \to \infty$. Because of the continuity of the map $t \mapsto \tau_{t,s} \circ \delta_t(B)$, (5.13) is verified on the dense set $\mathcal{U}_0 \subset \mathrm{Dom}(\delta_t)$.

To prove uniqueness, assume that $\{\hat{\tau}_{t,s}\}_{s,t\in\mathbb{R}}$ is any family of bounded operators on $\mathcal{U}$ satisfying (5.13) on $\mathcal{U}_0$. By (5.1) and because $\tau_{t,s}^{(L)}(B) \in \mathcal{U}_0$ for any $B \in \mathcal{U}_0$,

$$\hat{\tau}_{t,s}(B) - \tau_{t,s}^{(L)}(B) = \int_s^t \hat{\tau}_{\alpha,s} \circ \left(\delta_\alpha - \delta_\alpha^{(L)} \right) \circ \tau_{t,\alpha}^{(L)}(B) \, d\alpha \tag{5.15}$$

for any $B \in \mathcal{U}_0$, $L \in \mathbb{R}_0^+$ and $s, t \in \mathbb{R}$. Similar to (4.33)–(4.35), we infer from Theorem 5.1 (i) that, for any $\Lambda \in \mathcal{P}_f(\mathfrak{L})$, $B \in \mathcal{U}_\Lambda$, $\alpha, t \in \mathbb{R}$ and sufficiently large $L \in \mathbb{R}_0^+$,

$$\left\| \left(\delta_\alpha - \delta_\alpha^{(L)} \right) \circ \tau_{t,\alpha}^{(L)} (B) \right\|_{\mathcal{U}} \leq \| \Psi \|_\infty \, e^{2\mathbf{D}|t-\alpha| \| \Psi \|_\infty} \sum_{y \in \Lambda_L^c} \sum_{x \in \Lambda} \mathbf{F} \left(|x - y| \right) .$$

In particular, by (4.36), for any $B \in \mathcal{U}_0$ and $\alpha, t \in \mathbb{R}$,

$$\lim_{L \to \infty} \left\| \left(\delta_\alpha - \delta_\alpha^{(L)} \right) \circ \tau_{t,\alpha}^{(L)} (B) \right\|_{\mathcal{U}} = 0 \qquad (5.16)$$

uniformly for α on compacta. Because of (5.15) and $\{\hat{\tau}_{t,s}\}_{s,t \in \mathbb{R}} \subset \mathcal{B}(\mathcal{U})$, we then conclude from (5.16) that, for every $s, t \in \mathbb{R}$, $\hat{\tau}_{t,s}$ coincides on the dense set $\mathcal{U}_0$ with the limit $\tau_{t,s}$ of $\tau_{t,s}^{(L)}$, as $L \to \infty$. By continuity, $\tau_{t,s} = \hat{\tau}_{t,s}$ on $\mathcal{U}$ for any $s, t \in \mathbb{R}$. ∎

The solution of (5.13) exists under very weak conditions on interactions and potentials, i.e., their continuity, like in the finite-volume case. It yields a fundamental solution for the states of the interacting lattice fermions driven by the time-dependent interaction $\{\Psi^{(t)}\}_{t \in \mathbb{R}}$. More precisely, for any fixed $\rho_s \in \mathcal{U}^*$ at time $s \in \mathbb{R}$, the family $\{\rho_s \circ \tau_{t,s}\}_{t \in \mathbb{R}}$ solves the following ordinary differential equations, for each $B \in \mathcal{U}_0$:

$$\forall t \in \mathbb{R} : \qquad \partial_t \rho_t(B) = \rho_t \circ \delta_t(B) . \qquad (5.17)$$

By Corollary 5.2, the initial value problem on $\mathcal{U}^*$ associated with the above infinite system of ordinary differential equations is *well-posed*. Indeed, the solution of (5.17) is unique: Take any solution $\{\rho_t\}_{t \in \mathbb{R}}$ of (5.17) and, similar to (5.15), use the equality

$$\rho_t (B) - \rho_s \circ \tau_{t,s}^{(L)} (B) = \int_s^t \rho_\alpha \left(\left(\delta_\alpha - \delta_\alpha^{(L)} \right) \circ \tau_{t,\alpha}^{(L)} (B) \right) d\alpha$$

for any $\rho_s \in \mathcal{U}^*$, $B \in \mathcal{U}_0$, $L \in \mathbb{R}_0^+$ and $s, t \in \mathbb{R}$ together with (5.16) and the weak*-convergence of $\rho_s \circ \tau_{t,s}^{(L)}$ to $\rho_s \circ \tau_{t,s}$, as $L \to \infty$, by Corollary 5.2.

Note again that (5.13) is the non-autonomous evolution equation one formally obtains from the Schrödinger equation for automorphisms of the algebra of observables. See Sect. 2.4, in particular Eqs. (2.15) and (2.16). A similar remark can be done for the infinite system (5.17) of ordinary differential equations.

It is a priori unclear whether $\{\tau_{t,s}\}_{s,t \in \mathbb{R}}$ solves the non-autonomous Cauchy initial value problem

$$\forall s, t \in \mathbb{R} : \qquad \partial_s \tau_{t,s} = -\delta_s \circ \tau_{t,s} , \qquad \tau_{t,t} = \mathbf{1}_\mathcal{U} , \qquad (5.18)$$

on some dense domain. The generators $\{\delta_t\}_{t \in \mathbb{R}}$ are generally unbounded operators acting on $\mathcal{U}$ and their domains can additionally depend on time. As explained in Sect. 2.4, no unified theory of such linear evolution equations, similar to the Hille–Yosida generation theorems in the autonomous case, is available. See, e.g., [K4, C, S, P, BB] and the corresponding references therein.

By using Lieb–Robinson bounds for multi-commutators, we show below in Theorem 5.5 that the evolution equation (5.18) *also holds* on the dense set $\mathcal{U}_0$, under conditions like polynomial decays of interactions and boundedness of the external potential. Another example – more restrictive in which concerns the time-dependency of the generator of dynamics, but less restrictive w.r.t. the behavior at large distances of the potential $\mathbf{V}$ – for which (5.18) holds is given by Theorem 5.7 (i) in Sect. 5.3.

5.2 Lieb–Robinson Bounds for Multi-commutators

As explained in Remark 4.13, all results of Sect. 4.4 depend on Theorem 4.8 (iii). It is the crucial ingredient we need in order to prove Lemma 4.9, from which we derive Lieb–Robinson bounds for multi-commutators. Theorem 5.1 (ii) together with Corollary 5.2 extend Theorem 4.8 (iii) to time-dependent interactions and potentials. This allows us to prove Lemma 4.9 in the non-autonomous case as well. It is then straightforward to extend Lieb–Robinson bounds for multi-commutators to time-dependent interactions and potentials.

Recall that the proof of Lemma 4.9 uses that the space translated finite-volume groups $\{\tau_t^{(n,x)}\}_{t\in\mathbb{R}}$, $x \in \mathfrak{L}$, have all the same limit $\{\tau_t\}_{t\in\mathbb{R}}$, as $n \to \infty$. This also holds in the non-autonomous case. Indeed, for any $n \in \mathbb{N}_0$, $x \in \mathfrak{L}$, every bounded family $\Psi \doteq \{\Psi^{(t)}\}_{t\in\mathbb{R}}$ on $\mathcal{W}$ (i.e., $\|\Psi\|_\infty < \infty$), and each collection $\{\mathbf{V}^{(t)}\}_{t\in\mathbb{R}}$ of potentials, consider the (space) translated family $\{\tau_{t,s}^{(n,x)}\}_{s,t\in\mathbb{R}}$ of finite-volume $*$-automorphisms generated (cf. (5.1) and (5.2)) by the symmetric bounded derivation

$$\delta_t^{(n,x)}(B) \doteq i \sum_{\Lambda\subseteq\Lambda_n+x} \left[\Psi_\Lambda^{(t)}, B\right] + i \sum_{y\in\Lambda_n+x} \left[\mathbf{V}_{\{y\}}^{(t)}, B\right], \qquad B \in \mathcal{U}.$$

In the autonomous case the strong convergence of these evolution families towards $\{\tau_{t,s}\}_{s,t\in\mathbb{R}}$ easily follows from the second Trotter–Kato approximation theorem [EN, Chap. III, Sect. 4.9]. We use the Lieb–Robinson bound of Theorem 5.1 (i) to prove it in the non-autonomous case:

Lemma 5.3 (Limit of translated dynamics)
Let $\Psi \doteq \{\Psi^{(t)}\}_{t\in\mathbb{R}}$ *be a bounded family on* $\mathcal{W}$ *(i.e.,* $\|\Psi\|_\infty < \infty$*) and* $\{\mathbf{V}^{(t)}\}_{t\in\mathbb{R}}$ *a collection of potentials. For any* $y \in \mathfrak{L}$ *and* $\Lambda \in \mathcal{P}_f(\mathfrak{L})$*, assume* $\Psi_\Lambda, \mathbf{V}_{\{y\}} \in C(\mathbb{R}; \mathcal{U})$*. Then*

$$\lim_{n\to\infty} \tau_{t,s}^{(n,x)}(B) = \tau_{t,s}(B), \qquad B \in \mathcal{U}, \ x \in \mathfrak{L}, \ s, t \in \mathbb{R}.$$

Proof For any $n \in \mathbb{N}_0$ and $x \in \mathfrak{L}$, the translated finite-volume family $\{\tau_{s,t}^{(n,x)}\}_{s,t\in\mathbb{R}}$ solves non-autonomous evolution equations like (5.1) and (5.2). Therefore, similar to (5.15), for any $n \in \mathbb{N}_0$, $x \in \mathfrak{L}$, $\Lambda \in \mathcal{P}_f(\mathfrak{L})$, $B \in \mathcal{U}_\Lambda$ and $s, t \in \mathbb{R}$,

$$\tau_{t,s}^{(n,x)}(B) - \tau_{t,s}^{(n,0)}(B) = \int_s^t \tau_{\alpha,s}^{(n,x)} \circ \left(\delta_\alpha^{(n,x)} - \delta_\alpha^{(n,0)}\right) \circ \tau_{t,\alpha}^{(n,0)}(B)\,d\alpha\,. \qquad (5.19)$$

For sufficiently large $n \in \mathbb{N}_0$ such that $\Lambda \subset (\Lambda_n + x) \cap \Lambda_n$, note that

$$\left\|\left(\delta_\alpha^{(n,x)} - \delta_\alpha^{(n,0)}\right) \circ \tau_{t,\alpha}^{(n,0)}(B)\right\|_{\mathcal{U}} \leq \sum_{\mathcal{Z}\in\mathcal{P}_f(\mathfrak{L}),\ \mathcal{Z}\cap((\Lambda_n+x)^c\cup\Lambda_n^c)\neq\emptyset} \left\|\left[\Psi_\Lambda^{(t)}, \tau_{t,\alpha}^{(n,0)}(B)\right]\right\|_{\mathcal{U}}$$

with $\mathcal{Z}^c \doteq \mathfrak{L}\backslash\mathcal{Z}$ being the complement of any set $\mathcal{Z} \in \mathcal{P}_f(\mathfrak{L})$. Then, similar to Inequality (4.35), by using Theorem 5.1 (i), one verifies that, for any $x \in \mathfrak{L}$, $\Lambda \in \mathcal{P}_f(\mathfrak{L})$, $B \in \mathcal{U}_\Lambda$, $\alpha, t \in \mathbb{R}$, and sufficiently large $n \in \mathbb{N}_0$,

$$\left\|\left(\delta_\alpha^{(n,x)} - \delta_\alpha^{(n,0)}\right) \circ \tau_{t,\alpha}^{(n,0)}(B)\right\|_{\mathcal{U}} \qquad (5.20)$$
$$\leq 2\,\|B\|_{\mathcal{U}}\,\|\Psi\|_\infty\,e^{2\mathbf{D}|t-\alpha|\|\Psi\|_\infty} \sum_{y\in(\Lambda_n+x)^c\cup\Lambda_n^c} \sum_{z\in\Lambda} \mathbf{F}(|z-y|)\,,$$

while

$$\lim_{n\to\infty} \sum_{y\in(\Lambda_n+x)^c\cup\Lambda_n^c} \sum_{z\in\Lambda} \mathbf{F}(|z-y|) = 0\,, \qquad (5.21)$$

because of (4.5). We thus deduce from (5.20) and (5.21) that

$$\lim_{n\to\infty} \left\|\left(\delta_\alpha^{(n,x)} - \delta_\alpha^{(n,0)}\right) \circ \tau_{t,\alpha}^{(n,0)}(B)\right\|_{\mathcal{U}} = 0$$

uniformly for α on compacta. Combined with (5.19) and Corollary 5.2, this uniform limit implies the assertion. $\blacksquare$

With the above result and the introducing remarks of this subsection, it is now straightforward to extend Theorem 4.10 to the non-autonomous case:

Theorem 5.4 (Lieb–Robinson bounds for multi-commutators – Part I)
Let $\Psi \doteq \{\Psi^{(t)}\}_{t\in\mathbb{R}}$ *be a bounded family on* $\mathcal{W}$ *(i.e.,* $\|\Psi\|_\infty < \infty$*),* $\{\mathbf{V}^{(t)}\}_{t\in\mathbb{R}}$ *a collection of potentials, and* $s \in \mathbb{R}$*. For any* $y \in \mathfrak{L}$ *and* $\Lambda \in \mathcal{P}_f(\mathfrak{L})$*, assume* $\Psi_\Lambda, \mathbf{V}_{\{y\}} \in C(\mathbb{R};\mathcal{U})$*. Then, for any integer* $k \in \mathbb{N}$*,* $\{m_j\}_{j=0}^k \subset \mathbb{N}_0$*, times* $\{s_j\}_{j=1}^k \subset \mathbb{R}$*, lattice sites* $\{x_j\}_{j=0}^k \subset \mathfrak{L}$*, and elements* $B_0 \in \mathcal{U}_0$*,* $\{B_j\}_{j=1}^k \subset \mathcal{U}_0 \cap \mathcal{U}^+$ *such that* $B_j \in \mathcal{U}_{\Lambda_{m_j}}$ *for* $j \in \{0, \dots, k\}$*,*

$$\left\|\left[\tau_{s_k,s} \circ \chi_{x_k}(B_k), \dots, \tau_{s_1,s} \circ \chi_{x_1}(B_1), \chi_{x_0}(B_0)\right]^{(k+1)}\right\|_{\mathcal{U}}$$
$$\leq 2^k \prod_{j=0}^k \|B_j\|_{\mathcal{U}} \sum_{T\in\mathcal{T}_{k+1}} \left(\varkappa_T\left(\{(m_j, x_j)\}_{j=0}^k\right) + \mathfrak{R}_{T,\|\Psi\|_\infty}\right)\,,$$

where $\varkappa_T$ and $\mathfrak{R}_{T,\alpha}$ are respectively defined by (4.48) and (4.50) for $T \in \mathcal{T}_{k+1}$ and $\alpha \in \mathbb{R}_0^+$, the times $\{s_j\}_{j=1}^k$ in (4.50) being replaced with $\{(s_j - s)\}_{j=1}^k$.

Proof One easily checks that Theorem 5.1 (ii) holds for $\{\tau_{t,s}^{(n,x)}\}_{s,t\in\mathbb{R}}$ at any fixed $x \in \mathfrak{L}$ and $n \in \mathbb{N}_0$. By Lemmas 5.3 and 4.9 also holds in the non-autonomous case and the assertion follows from (4.54) with the $*$-automorphism τ_{s_j} being replaced by $\tau_{s_j,s}$ for every $j \in \{1, \ldots, k\}$. ∎

By Theorems 4.11 and 5.4, we obtain Lieb–Robinson bounds for multi-commutators as well as a version of Corollary 4.12 in the *non-autonomous* case. I.e., interacting and non-autonomous systems also satisfy the so-called tree-decay bounds.

Another application of Theorems 4.11 and 5.4 is a proof of existence of a fundamental solution for the non-autonomous abstract Cauchy initial value problem for observables

$$\forall s \in \mathbb{R}: \quad \partial_s B_s = -\delta_s(B_s), \quad B_t = B \in \mathcal{U}_0, \quad (5.22)$$

in the Banach space $\mathcal{U}$, i.e., a proof of existence of a solution of the evolution equation (5.18). The latter is a non-trivial statement, as previously discussed, among other things because the domain of δ_s depends, in general, on the time $s \in \mathbb{R}$. [Here, $t \in \mathbb{R}$ is the "initial" time.]

To this end, like in (4.77)–(4.80), we add the following condition on interactions Φ:

- *Polynomial decay.* Assume (4.56) and the existence of constants $\upsilon, D \in \mathbb{R}^+$ such that

$$\sup_{x \in \mathfrak{L}} \sum_{\Lambda \in \mathcal{D}(x,m)} \|\Phi_\Lambda\|_{\mathcal{U}} \leq D(m+1)^{-\upsilon}, \quad m \in \mathbb{N}_0, \quad (5.23)$$

while the sequence $\{\mathbf{u}_{n,m}\}_{n\in\mathbb{N}} \in \ell^1(\mathbb{N})$ of (4.56) satisfies

$$\sum_{m,n\in\mathbb{N}} m^{-\upsilon} |\mathbf{u}_{n,m}| < \infty. \quad (5.24)$$

As $\mathbf{F}(|x|) > 0$ for all $x \in \mathfrak{L}$, note that (4.56) implies

$$\sum_{n\in\mathbb{N}} |\mathbf{u}_{n,m}| \geq Dm^\varsigma$$

for some $D \in \mathbb{R}^+$ and all $m \in \mathbb{N}_0$. Hence, the inequality (5.24) imposes $\upsilon > \varsigma + 1$.
Then, one gets the following assertion:

Theorem 5.5 (Dynamics and non-autonomous evolution equations)
Let $\Psi \doteq \{\Psi^{(t)}\}_{t\in\mathbb{R}} \in C(\mathbb{R}; \mathcal{W})$ be a bounded family on $\mathcal{W}$ (i.e., $\|\Psi\|_\infty < \infty$) and $\{\mathbf{V}_{\{x\}}^{(t)}\}_{x\in\mathfrak{L},t\in\mathbb{R}}$ a bounded family on $\mathcal{U}$ of potentials with $\mathbf{V}_{\{x\}} \in C(\mathbb{R}; \mathcal{U})$ for any $x \in \mathfrak{L}$. Assume (4.56) with $\varsigma > 2d$ and that (5.23) and (5.24) with $\Phi = \Psi^{(t)}$ and

$\nu > \varsigma + 1$ *hold uniformly for* $t \in \mathbb{R}$. *Then, for any* $s, t \in \mathbb{R}$, $\tau_{t,s}(\mathcal{U}_0) \subset \mathrm{Dom}(\delta_s)$ *and* $\{\tau_{t,s}\}_{s,t \in \mathbb{R}}$ *solves the non-autonomous evolution equation*

$$\forall s, t \in \mathbb{R}: \quad \partial_s \tau_{t,s} = -\delta_s \circ \tau_{t,s}\,, \quad \tau_{t,t} = \mathbf{1}_{\mathcal{U}}\,, \tag{5.25}$$

in the strong sense on the dense set $\mathcal{U}_0$.

Proof 1. Let $s, t \in \mathbb{R}$, $\Lambda \in \mathcal{P}_f(\mathfrak{L})$ and take any element $B \in \mathcal{U}_\Lambda$. As a preliminary step, we prove that $\{\delta_s \circ \tau_{t,s}^{(L)}(B)\}_{L \in \mathbb{R}_0^+}$ converges to $\delta_s \circ \tau_{t,s}(B)$, as $L \to \infty$. In particular, $\tau_{t,s}(\mathcal{U}_0) \subset \mathrm{Dom}(\delta_s)$. By using similar arguments as in the proof of Theorem 5.1 (ii), it suffices to study the limit of $\{\delta_s \circ \tilde{\tau}_{t,s}^{(L)}(B)\}_{L \in \mathbb{R}_0^+}$, see (5.10).

Similar to (4.23), from (5.6)–(5.11) and straightforward computations, for any $L_1, L_2 \in \mathbb{R}_0^+$ with $\Lambda \subset \Lambda_{L_1} \subsetneq \Lambda_{L_2}$,

$$\left\| \delta_s \circ \left(\tilde{\tau}_{t,s}^{(L_2)}(B) - \tilde{\tau}_{t,s}^{(L_1)}(B) \right) \right\|_{\mathcal{U}} \tag{5.26}$$
$$\leq \int_{\min\{s,t\}}^{\max\{s,t\}} \sum_{\mathcal{Z} \in \mathcal{P}_f(\mathfrak{L})} \left\| \left[\hat{\tau}_{s,\alpha}^{(L_1,L_2)}(\Psi_{\mathcal{Z}}^{(s)}), B_\alpha^{(L_1,L_2)}, \tau_{t,\alpha}^{(L_1)}(\tilde{B}_t) \right]^{(3)} \right\|_{\mathcal{U}} d\alpha\,,$$

where $\tilde{B}_t \doteq \mathcal{V}_{t,s}(\mathbf{V}_\Lambda) B \mathcal{V}_{s,t}(\mathbf{V}_\Lambda)$,

$$\hat{\tau}_{s,\alpha}^{(L_1,L_2)}(B) \doteq \mathcal{V}_{s,\alpha}(\mathbf{V}_{\Lambda_{L_2} \setminus \Lambda_{L_1}}) \tau_{s,\alpha}^{(L_2)}(B) \mathcal{V}_{\alpha,s}(\mathbf{V}_{\Lambda_{L_2} \setminus \Lambda_{L_1}})\,, \quad B \in \mathcal{U}, \ s, \alpha \in \mathbb{R}\,, \tag{5.27}$$

and

$$B_\alpha^{(L_1,L_2)} \doteq \sum_{\mathcal{Z} \subseteq \Lambda_{L_2},\ \mathcal{Z} \cap (\Lambda_{L_2} \setminus \Lambda_{L_1}) \neq \emptyset} \mathcal{V}_{\alpha,s}(\mathbf{V}_{\Lambda_{L_2} \setminus \Lambda_{L_1}}) \Psi_{\mathcal{Z}} \mathcal{V}_{s,\alpha}(\mathbf{V}_{\Lambda_{L_2} \setminus \Lambda_{L_1}}) \in \mathcal{U}^+ \cap \mathcal{U}_{\Lambda_{L_2}}\,.$$

Using (5.12), observe that, for all $\mathcal{Z} \subseteq \Lambda_{L_2}$ and $\alpha, s \in \mathbb{R}$,

$$\mathcal{V}_{\alpha,s}(\mathbf{V}_{\Lambda_{L_2} \setminus \Lambda_{L_1}}) \Psi_{\mathcal{Z}} \mathcal{V}_{s,\alpha}(\mathbf{V}_{\Lambda_{L_2} \setminus \Lambda_{L_1}}) \in \mathcal{U}^+ \cap \mathcal{U}_{\mathcal{Z}} \tag{5.28}$$

with

$$\left\| \mathcal{V}_{\alpha,s}(\mathbf{V}_{\Lambda_{L_2} \setminus \Lambda_{L_1}}) \Psi_{\mathcal{Z}} \mathcal{V}_{s,\alpha}(\mathbf{V}_{\Lambda_{L_2} \setminus \Lambda_{L_1}}) \right\|_{\mathcal{U}} = \| \Psi_{\mathcal{Z}} \|_{\mathcal{U}}\,. \tag{5.29}$$

Similarly, for all $t \in \mathbb{R}$,

$$\tilde{B}_t \in \mathcal{U}_\Lambda \quad \text{and} \quad \| \tilde{B}_t \|_{\mathcal{U}} = \| B \|_{\mathcal{U}}\,. \tag{5.30}$$

In order to bound the sum

$$\sum_{\mathcal{Z} \in \mathcal{P}_f(\mathfrak{L})} \left[\hat{\tau}_{s,\alpha}^{(L_1,L_2)}(\Psi_{\mathcal{Z}}^{(s)}), B_\alpha^{(L_1,L_2)}, \tau_{t,\alpha}^{(L_1)}(\tilde{B}_t) \right]^{(3)} \tag{5.31}$$

of multi-commutators of order three we represent it as a convenient series, whose summability is uniform w.r.t. $L_1, L_2 \in \mathbb{R}_0^+$ ($\Lambda \subset \Lambda_{L_1} \subsetneq \Lambda_{L_2}$). To this end, first develop $\tau_{t,\alpha}^{(L_1)}(\tilde{B}_t)$ as a telescoping series: Let $m_0 \in \mathbb{N}_0$ be the smallest integer such that $\Lambda \subset \Lambda_{m_0}$. Then, similar to Lemma 4.9 (autonomous case) and as explained in the proof of Theorem 5.4, for any $\alpha, t \in \mathbb{R}$ and $L_1 \in \mathbb{R}_0^+$,

$$\tau_{t,\alpha}^{(L_1)}(\tilde{B}_t) = \sum_{n=m_0}^{\infty} \tilde{\mathfrak{B}}_{t,\alpha}(n) .$$

Here, for all integers $n \geq m_0$, $\tilde{\mathfrak{B}}_{t,\alpha}(n) \in \mathcal{U}_{\Lambda_n}$ where $\|\tilde{\mathfrak{B}}_{t,\alpha}(m_0)\|_{\mathcal{U}} = \|B\|_{\mathcal{U}}$ (see (5.30)) and, for all $n \in \mathbb{N}$ with $n > m_0$,

$$\|\tilde{\mathfrak{B}}_{t,\alpha}(n)\|_{\mathcal{U}} \leq 2\|B\|_{\mathcal{U}} \|\Psi\|_\infty |t - \alpha| \, e^{4\mathbf{D}|t-\alpha|\|\Psi\|_\infty} \frac{\mathbf{u}_{n,m_0}}{(1+n)^\varsigma} , \tag{5.32}$$

by Theorem 5.1 (ii) and Assumption (4.56). Of course, $\tilde{\mathfrak{B}}_{t,\alpha}(n) = 0$ for any integer $n > L_1$ and $\alpha, t \in \mathbb{R}$ because $\{\tau_{t,s}^{(L_1)}\}_{s,t \in \mathbb{R}}$ is a finite-volume dynamics. Meanwhile, because of (5.12), Theorem 5.4 holds by replacing $\{\tau_{t,s}\}_{s,t \in \mathbb{R}}$ with $\{\hat{\tau}_{t,s}^{(L_1,L_2)}\}_{s,t \in \mathbb{R}}$ at sufficiently large $L_1, L_2 \in \mathbb{R}_0^+$ ($\Lambda_{L_1} \subsetneq \Lambda_{L_2}$). Using this together with (5.23) and (5.24) for $\Phi = \Psi^{(t)}$, Eqs. (5.28)–(5.30), Theorem 4.11, as well as the assumptions $\nu > \varsigma + 1$ and $\varsigma > 2d$,

$$\sum_{n_0=m_0}^{\infty} \sum_{x_2 \in \mathfrak{L}} \sum_{m_2 \in \mathbb{N}_0} \sum_{\mathcal{Z}_2 \in \mathcal{D}(x_2,m)} \sum_{x_1 \in \mathfrak{L}} \sum_{m_1 \in \mathbb{N}_0} \sum_{\mathcal{Z}_1 \in \mathcal{D}(x_1,m)} \tag{5.33}$$

$$\left\| \left[\hat{\tau}_{s,\alpha}^{(L_1,L_2)}(\Psi_{\mathcal{Z}_2}^{(s)}), \mathcal{V}_{\alpha,s}(\mathbf{V}_{\Lambda_{L_2}\backslash\Lambda_{L_1}})\Psi_{\mathcal{Z}_1}\mathcal{V}_{s,\alpha}(\mathbf{V}_{\Lambda_{L_2}\backslash\Lambda_{L_1}}), \tilde{\mathfrak{B}}_{t,\alpha}(n) \right]^{(3)} \right\|_{\mathcal{U}}$$

$$\leq \mathbf{D} \|B\|_{\mathcal{U}} \|\mathbf{u}_{\cdot,m_0}\|_{\ell^1(\mathbb{N})} \left(\sum_{m_1 \in \mathbb{N}_0} (m_1+1)^{\varsigma - \nu} \right)$$

$$\times \sum_{m_2 \in \mathbb{N}_0} (m_2+1)^{-\nu} \left(\sum_{n_2 \in \mathbb{N}} \mathbf{u}_{n_2,m_2} + (m_2+1)^\varsigma \right) < \infty .$$

Similar to (5.32) and because (5.23) and (5.24) with $\Phi = \Psi^{(t)}$ hold uniformly for $t \in \mathbb{R}$, the strictly positive constant $\mathbf{D} \in \mathbb{R}^+$ is uniformly bounded for s, t, α on compacta and $L_1, L_2 \in \mathbb{R}_0^+$ ($\Lambda \subset \Lambda_{L_1} \subsetneq \Lambda_{L_2}$). The last sum is an *upper bound* of the integrand of the r.h.s. of (5.26). Indeed, we deduce from (4.79) that

$$B_\alpha^{(L_1,L_2)} = \sum_{x \in \Lambda_{L_2}\backslash\Lambda_{L_1}} \sum_{m \in \mathbb{N}_0} \sum_{\mathcal{Z} \subseteq \Lambda_{L_2}, \, \mathcal{Z} \in \mathcal{D}(x,m)}$$

$$\frac{1}{|\mathcal{Z} \cap \Lambda_{L_2}\backslash\Lambda_{L_1}|} \mathcal{V}_{\alpha,s}(\mathbf{V}_{\Lambda_{L_2}\backslash\Lambda_{L_1}})\Psi_{\mathcal{Z}}\mathcal{V}_{s,\alpha}(\mathbf{V}_{\Lambda_{L_2}\backslash\Lambda_{L_1}})$$

and

$$\sum_{\mathcal{Z}\in\mathcal{P}_f(\mathfrak{L}),\ \mathcal{Z}\cap\Lambda_{L_2}\neq\emptyset}\hat{\tau}_{s,\alpha}^{(L_1,L_2)}(\Psi_{\mathcal{Z}}^{(s)})=\sum_{x\in\Lambda_{L_2}}\sum_{m\in\mathbb{N}_0}\sum_{\mathcal{Z}\in\mathcal{D}(x,m)}\frac{1}{|\mathcal{Z}\cap\Lambda_{L_2}|}\hat{\tau}_{s,\alpha}^{(L_1,L_2)}(\Psi_{\mathcal{Z}}^{(s)}).$$

[Compare this last sum with (5.31) by using (5.28) and (5.30) to restrict the whole sum over $\mathcal{Z}\in\mathcal{P}_f(\mathfrak{L})$ to finite sets $\mathcal{Z}$ so that $\mathcal{Z}\cap\Lambda_{L_2}\neq\emptyset$.]

As a consequence, for any $s,t\in\mathbb{R}$ and $B\in\mathcal{U}_0$, we infer from (5.26), (5.33), and Lebesgue's dominated convergence theorem that $\{\delta_s\circ\tilde{\tau}_{t,s}^{(L)}(B)\}_{L\in\mathbb{R}_0^+}$, and hence $\{\delta_s\circ\tau_{t,s}^{(L)}(B)\}_{L\in\mathbb{R}_0^+}$, are Cauchy nets within the complete space $\mathcal{U}$. By Corollary 5.2, $\{\tau_{t,s}^{(L)}\}_{L\in\mathbb{R}_0^+}$ converges strongly to $\tau_{t,s}$ for every $s,t\in\mathbb{R}$. Recall meanwhile that the operator δ_s is the *closed* operator described in Theorem 4.8 for the interaction $\Psi^{(s)}\in\mathcal{W}$ and the potential $\mathbf{V}^{(s)}$ at fixed $s\in\mathbb{R}$. Therefore, $\tau_{t,s}(B)\in\mathrm{Dom}(\delta_s)$ and the family $\{\delta_s\circ\tau_{t,s}^{(L)}(B)\}_{L\in\mathbb{R}_0^+}$ converges to $\delta_s\circ\tau_{t,s}(B)$, i.e.,

$$\lim_{L\to\infty}\left\|\delta_s\circ\left(\tau_{t,s}(B)-\tau_{t,s}^{(L)}(B)\right)\right\|_{\mathcal{U}}=0.\tag{5.34}$$

In particular, $\tau_{t,s}(\mathcal{U}_0)\subset\mathrm{Dom}(\delta_s)$.

Now, by using (5.1) one gets that, for $L\in\mathbb{R}_0^+$, $s,t,h\in\mathbb{R}$, $h\neq 0$, and $B\in\mathcal{U}_0$,

$$\begin{aligned}
&\left\||h|^{-1}\left(\tau_{t,s+h}(B)-\tau_{t,s}(B)\right)+\delta_s\circ\tau_{t,s}(B)\right\|_{\mathcal{U}}\\
&\leq\left\|\delta_s\circ\left(\tau_{t,s}(B)-\tau_{t,s}^{(L)}(B)\right)\right\|_{\mathcal{U}}\\
&\quad+\sup_{\alpha\in[s-|h|,s+|h|]}\left\|\delta_s^{(L)}\circ\tau_{t,s}^{(L)}(B)-\delta_\alpha^{(L)}\circ\tau_{t,\alpha}^{(L)}(B)\right\|_{\mathcal{U}}\\
&\quad+\left\|\left(\delta_s^{(L)}-\delta_s\right)\circ\tau_{t,s}^{(L)}(B)\right\|_{\mathcal{U}}\\
&\quad+2|h|^{-1}\sup_{\alpha\in[s-|h|,s+|h|]}\left\|\tau_{t,\alpha}(B)-\tau_{t,\alpha}^{(L)}(B)\right\|_{\mathcal{U}}.
\end{aligned}\tag{5.35}$$

We proceed by estimating the four terms in the upper bound of (5.35). The first one is already analyzed, see (5.34). So, we start with the second. If nothing is explicitly mentioned, the parameters $L\in\mathbb{R}_0^+$, $s,t,h\in\mathbb{R}$, $\Lambda\in\mathcal{P}_f(\mathfrak{L})$ and $B\in\mathcal{U}_\Lambda$ are fixed.

2. For any $\alpha\in\mathbb{R}$, observe that

$$\begin{aligned}
\left\|\delta_s^{(L)}\circ\tau_{t,s}^{(L)}(B)-\delta_\alpha^{(L)}\circ\tau_{t,\alpha}^{(L)}(B)\right\|_{\mathcal{U}}&\leq\left\|\left(\delta_s^{(L)}-\delta_\alpha^{(L)}\right)\circ\tau_{t,\alpha}^{(L)}(B)\right\|_{\mathcal{U}}\\
&\quad+\left\|\delta_s^{(L)}\circ\left(\tau_{t,s}^{(L)}-\tau_{t,\alpha}^{(L)}\right)(B)\right\|_{\mathcal{U}}.
\end{aligned}\tag{5.36}$$

By using first (4.27) for the interaction $\Psi^{(s)}$ and potential $\mathbf{V}^{(s)}$ and then Lieb–Robinson bounds (Theorem 5.1 (i)) in the same way as (4.35), one verifies that, for any $\alpha\in\mathbb{R}$ and $B\neq 0$,

$$\frac{\left\| \left(\delta_s^{(L)} - \delta_\alpha^{(L)} \right) \circ \tau_{t,\alpha}^{(L)} (B) \right\|_{\mathcal{U}}}{2 \, \|B\|_{\mathcal{U}}} \qquad (5.37)$$

$$\leq \left\| \Psi^{(s)} - \Psi^{(\alpha)} \right\|_{\mathcal{W}} e^{2\mathbf{D}|t-\alpha| \|\Psi\|_\infty} |\Lambda| \, \|\mathbf{F}\|_{1,\mathfrak{L}} + \sum_{x \in \Lambda} \left\| \mathbf{V}_{\{x\}}^{(\alpha)} - \mathbf{V}_{\{x\}}^{(s)} \right\|_{\mathcal{U}}$$

$$+ \mathbf{D}^{+1} \left(e^{2\mathbf{D}|t-\alpha| \|\Psi\|_\infty} - 1 \right) \sum_{x \in \mathfrak{L} \setminus \Lambda} \left\| \mathbf{V}_{\{x\}}^{(\alpha)} - \mathbf{V}_{\{x\}}^{(s)} \right\|_{\mathcal{U}} \sum_{y \in \Lambda} \mathbf{F} \left(|x - y| \right) .$$

By assumption, $\Psi \in C(\mathbb{R}; \mathcal{W})$, $\{ \mathbf{V}_{\{x\}}^{(t)} \}_{x \in \mathfrak{L}, t \in \mathbb{R}}$ is a bounded family in $\mathcal{U}$, and $\mathbf{V}_{\{x\}} \in C(\mathbb{R}; \mathcal{U})$ for any $x \in \mathfrak{L}$. So, by Lebesgue's dominated convergence theorem, it follows from (5.37) that

$$\lim_{h \to 0} \sup_{\alpha \in [s - |h|, s + |h|]} \left\| \left(\delta_s^{(L)} - \delta_\alpha^{(L)} \right) \circ \tau_{t,\alpha}^{(L)} (B) \right\|_{\mathcal{U}} = 0 . \qquad (5.38)$$

On the other hand, by (5.1),

$$\sup_{\alpha \in [s - |h|, s + |h|]} \left\| \delta_s^{(L)} \circ \left(\tau_{t,s}^{(L)} - \tau_{t,\alpha}^{(L)} \right) (B) \right\|_{\mathcal{U}} \leq \int_{s - |h|}^{s + |h|} \left\| \delta_s^{(L)} \circ \delta_\alpha^{(L)} \circ \tau_{t,\alpha}^{(L)} (B) \right\|_{\mathcal{U}} d\alpha , \qquad (5.39)$$

where

$$\left\| \delta_s^{(L)} \circ \delta_\alpha^{(L)} \circ \tau_{t,\alpha}^{(L)} (B) \right\|_{\mathcal{U}} \leq \sum_{\mathcal{Z}_1, \mathcal{Z}_2 \in \mathcal{P}_f(\mathfrak{L})} \left\| \left[\Psi_{\mathcal{Z}_1}^{(s)}, \Psi_{\mathcal{Z}_2}^{(\alpha)}, \tau_{t,\alpha}^{(L)} (B) \right]^{(3)} \right\|_{\mathcal{U}} \qquad (5.40)$$

$$+ \sum_{\mathcal{Z} \in \mathcal{P}_f(\mathfrak{L})} \sum_{x \in \mathfrak{L}} \left\| \left[\Psi_{\mathcal{Z}}^{(s)}, \mathbf{V}_{\{x\}}^{(\alpha)}, \tau_{t,\alpha}^{(L)} (B) \right]^{(3)} \right\|_{\mathcal{U}}$$

$$+ \sum_{\mathcal{Z} \in \mathcal{P}_f(\mathfrak{L})} \sum_{x \in \mathfrak{L}} \left\| \left[\mathbf{V}_{\{x\}}^{(s)}, \Psi_{\mathcal{Z}}^{(\alpha)}, \tau_{t,\alpha}^{(L)} (B) \right]^{(3)} \right\|_{\mathcal{U}}$$

$$+ \sum_{x, y \in \mathfrak{L}} \left\| \left[\mathbf{V}_{\{x\}}^{(s)}, \mathbf{V}_{\{y\}}^{(\alpha)}, \tau_{t,\alpha}^{(L)} (B) \right]^{(3)} \right\|_{\mathcal{U}} .$$

Similar to (5.33), we use Theorems 4.11 (i) and 5.4 for $k = 2$ to derive an upper bound for the r.h.s. of (5.40), uniformly w.r.t. large $L \in \mathbb{R}_0^+$ and $\alpha \in [s - 1, s + 1]$. By (5.39), it follows that

$$\lim_{h \to 0} \sup_{\alpha \in [s - |h|, s + |h|]} \left\| \delta_s^{(L)} \circ \left(\tau_{t,s}^{(L)} - \tau_{t,\alpha}^{(L)} \right) (B) \right\|_{\mathcal{U}} = 0 .$$

Combined with (5.36) and (5.38) this yields

$$\lim_{h \to 0} \sup_{\alpha \in [s - |h|, s + |h|]} \left\| \delta_s^{(L)} \circ \tau_{t,s}^{(L)} (B) - \delta_\alpha^{(L)} \circ \tau_{t,\alpha}^{(L)} (B) \right\|_{\mathcal{U}} = 0 . \qquad (5.41)$$

3. Similar to (4.35), one gets from Lieb–Robinson bounds (Theorem 5.1 (i)) that

$$\left\| \left(\delta_s^{(L)} - \delta_s \right) \circ \tau_{t,s}^{(L)} (B) \right\|_{\mathcal{U}} \leq \|\Psi\|_\infty \, e^{2\mathbf{D}|t-s|\|\Psi\|_\infty} \sum_{y \in \Lambda_L^c} \sum_{x \in \Lambda} \mathbf{F}\left(|x - y| \right) ,$$

which combined with (4.36) gives

$$\lim_{L \to \infty} \left\| \left(\delta_s^{(L)} - \delta_s \right) \circ \tau_{t,s}^{(L)} (B) \right\|_{\mathcal{U}} = 0 . \tag{5.42}$$

4. In the limit $h \to 0$, we take $L_h \to \infty$ such that

$$\lim_{h \to 0} |h|^{-1} \sup_{\alpha \in [s-|h|, s+|h|]} \left\| \tau_{t,\alpha} (B) - \tau_{t,\alpha}^{(L_h)} (B) \right\|_{\mathcal{U}} = 0 . \tag{5.43}$$

This is possible because $\tau_{t,s}^{(L)} (B)$ converges to $\tau_{t,s} (B)$, uniformly for t, s on compacta, by Corollary 5.2. We eventually combine (5.34)–(5.43) with Inequality (5.35) to arrive at the assertion. ∎

Note that uniqueness of the solution of the non-autonomous evolution equation (5.25) cannot be proven as done for the proof of uniqueness in Corollary 5.2 (iii). Indeed, take any family $\{\hat{\tau}_{t,s}\}_{s,t \in \mathbb{R}}$ of bounded operators on $\mathcal{U}$ satisfying (5.25) on $\mathcal{U}_0$. Then, as before in the proof of Corollary 5.2 (iii), for any $B \in \mathcal{U}_0$, $L \in \mathbb{R}_0^+$ and $s, t \in \mathbb{R}$,

$$\tau_{t,s}^{(L)} (B) - \hat{\tau}_{t,s} (B) = \int_s^t \tau_{\alpha,s}^{(L)} \circ \left(\delta_\alpha^{(L)} - \delta_\alpha \right) \circ \hat{\tau}_{t,\alpha} (B) \, d\alpha , \tag{5.44}$$

by using (5.2). However, it is not clear this time whether the norm

$$\left\| \tau_{\alpha,s}^{(L)} \circ \left(\delta_\alpha^{(L)} - \delta_\alpha \right) \circ \hat{\tau}_{t,\alpha} (B) \right\|_{\mathcal{U}} = \left\| \left(\delta_\alpha - \delta_\alpha^{(L)} \right) \circ \hat{\tau}_{t,\alpha} (B) \right\|_{\mathcal{U}}$$

vanishes, as $L \to \infty$, even if (4.29) for δ_α and $\delta_\alpha^{(L)}$ holds true, because $\hat{\tau}_{t,\alpha} (B) \in$ Dom (δ_α) can be *outside* $\mathcal{U}_0$. The strong convergence of $\delta_\alpha^{(L)}$ to δ_α on some core of δ_α does not imply, in general, the strong convergence on any core of δ_α. The equality (5.44) with $\tau_{t,s}, \delta_s$ replacing $\tau_{t,s}^{(L)}, \delta_s^{(L)}$ is also not clear because (5.13) is only known to hold true on $\mathcal{U}_0$ and a priori *not* on the whole domain Dom (δ_α) of δ_α.

The non-autonomous evolution equation (5.22) of Theorem 5.5 is not parabolic because the symmetric derivation δ_t, $t \in \mathbb{R}$, is generally *not* the generator of an analytic semigroup. Note also that no Hölder continuity condition is imposed on $\{\delta_t\}_{t \in \mathbb{R}}$, like in the class of parabolic evolution equations introduced in [AT, Hypotheses I-II]. See also [S] or [P, Sect. 5.6.] for more simplified studies.

In fact, (5.22) is rather related to Kato's *hyperbolic* evolution equations [K2, K3, K4]. The so-called *Kato quasi-stability* is satisfied by the family of generators $\{\delta_t\}_{t \in \mathbb{R}}$ because they are always dissipative operators, by Lemma 4.5. $\{\delta_t\}_{t \in \mathbb{R}}$ is also strongly continuous on the dense set $\mathcal{U}_0$, which is a common core of all δ_t, $t \in \mathbb{R}$. However, in general, even for *finite* range interactions $\Psi \in \mathcal{W}$, the strongly continuous two-parameter family $\{\tau_{t,s}\}_{s,t\in\mathbb{R}}$ does *not* conserve the dense set $\mathcal{U}_0$, i.e., $\tau_{t,s}(\mathcal{U}_0) \not\subseteq \mathcal{U}_0$ for any $s \neq t$. In some specific situations one can directly show that the completion of the core $\mathcal{U}_0$ w.r.t. a conveniently chosen norm defines a so-called admissible Banach space $\mathcal{Y} \supset \mathcal{U}_0$ of the generator at any time, which satisfies further technical conditions leading to Kato's hyperbolic conditions [K2, K3, K4]. See also [P, Sect. 5.3.] and [BB, Sect. VII.1], which is used in the proof of Theorem 5.7 (i). Nevertheless, the existence of such a Banach space $\mathcal{Y}$ is a priori unclear in the general case treated in Theorem 5.5. See for instance the uniqueness problem explained just above.

Note that we only assume in Theorem 5.5 some polynomial decay for the interaction with (4.7) and (5.23)–(5.24) (uniformly in time). Recall that these assumptions are fulfilled for any interaction $\Psi \in \mathcal{W}$ with (4.7), provided the parameter $\epsilon \in \mathbb{R}^+$ is sufficiently large. In the case of exponential decays, stronger results can be deduced from Lieb–Robinson bounds for multi-commutators. For the interested reader, we give below one example, which is based on interactions Φ satisfying the following condition:

- *Exponential decay.* Assume (4.57) and the existence of constants $\upsilon > \varsigma$ and $D \in \mathbb{R}^+$ such that

$$\sup_{x\in\mathfrak{L}} \sum_{\Lambda\in\mathcal{D}(x,m)} \|\Phi_\Lambda\|_\mathcal{U} \leq De^{-\upsilon m}, \qquad m \in \mathbb{N}_0, \tag{5.45}$$

while

$$\sum_{m\in\mathbb{N}} \mathbf{C}_m e^{-(\varsigma+\upsilon)m} < \infty. \tag{5.46}$$

Theorem 5.6 (Graph norm convergence and Gevrey vectors)
Let $\Psi \doteq \{\Psi^{(t)}\}_{t\in\mathbb{R}}$ be a bounded family on $\mathcal{W}$ (i.e., $\|\Psi\|_\infty < \infty$), $\{\mathbf{V}^{(t)}\}_{t\in\mathbb{R}}$ a collection of potentials, and $B \in \mathcal{U}_0$. For any $x \in \mathfrak{L}$ and $\Lambda \in \mathcal{P}_f(\mathfrak{L})$, $\Psi_\Lambda, \mathbf{V}_{\{x\}} \in C(\mathbb{R}; \mathcal{U})$. Assume that (4.57) and (5.45)–(5.46) hold for $\Phi = \Psi^{(t)}$, uniformly in time.
(i) Graph norm convergence. As $L \to \infty$, $\tau_{t,s}^{(L)}(B)$ converges, uniformly for s, t on compacta, to $\tau_{t,s}(B)$ within the normed space $(\mathrm{Dom}(\delta_s^m), \|\cdot\|_{\delta_s^m})$, where, for all $m \in \mathbb{N}_0$, $\|\cdot\|_{\delta_s^m}$ stands for the graph norm of the densely defined operator δ_s^m.

(ii) Gevrey vectors. If $\{\mathbf{V}_{\{x\}}^{(t)}\}_{x\in\mathfrak{L},t\in\mathbb{R}}$ is a bounded family on $\mathcal{U}$ then, for any $\mathrm{T} \in \mathbb{R}_0^+$, there exist $r \equiv r_{d,\mathrm{T},\Psi,\mathbf{V},\mathbf{F}} \in \mathbb{R}^+$ and $D \equiv D_{\mathrm{T},\Psi,\mathbf{V}} \in \mathbb{R}^+$ such that, for all $s, t \in [-\mathrm{T}, \mathrm{T}]$, $m_0 \in \mathbb{N}_0$ and $B \in \mathcal{U}_{\Lambda_{m_0}}$,

$$\sum_{m\in\mathbb{N}} \frac{r^m}{(m!)^d} \|\delta_s^m \circ \tau_{t,s}(B)\|_\mathcal{U} \leq De^{m_0\varsigma} \|B\|_\mathcal{U}.$$

Proof (i) The case $m = 0$ follows from Corollary 5.2. Let $m \in \mathbb{N}$ and $B \in \mathcal{U}_0$. Similar to (5.26), for any sufficiently large $L_1, L_2 \in \mathbb{R}_0^+$, $\Lambda_{L_1} \subsetneq \Lambda_{L_2}$,

$$\left\| \delta_s^m \circ \left(\tilde{\tau}_{t,s}^{(L_2)}(B) - \tilde{\tau}_{t,s}^{(L_1)}(B) \right) \right\|_{\mathcal{U}}$$

$$\leq \int_{\min\{s,t\}}^{\max\{s,t\}} \sum_{\mathcal{Z}_1,\ldots,\mathcal{Z}_m \in \mathcal{P}_f(\mathfrak{L})} \left\| \left[\hat{\tau}_{s,\alpha}^{(L_1,L_2)}(\Psi_{\mathcal{Z}_m}^{(s)}), \ldots, \hat{\tau}_{s,\alpha}^{(L_1,L_2)}(\Psi_{\mathcal{Z}_1}^{(s)}), \right. \right.$$

$$\left. \left. , B_\alpha^{(L_1,L_2)}, \tau_{t,\alpha}^{(L_1)}(\tilde{B}_t) \right]^{(m+2)} \right\|_{\mathcal{U}} d\alpha \,, \tag{5.47}$$

see (5.27). From a straightforward generalization of (5.33) for multi-commutators of degree $m + 2$ and the same kind of arguments used in point **1.** of the proof of Theorem 5.5, the r.h.s. of the above inequality tends to zero in the limit of large $L_1, L_2 \in \mathbb{R}_0^+$ ($\Lambda_{L_1} \subsetneq \Lambda_{L_2}$). This holds for every $m \in \mathbb{N}$ because the interaction has, by assumption, exponential decay, see (4.57) and (5.45)–(5.46).

Consequently, $\{\delta_s^m \circ \tilde{\tau}_{t,s}^{(L)}(B)\}_{L \in \mathbb{R}_0^+}$, and hence $\{\delta_s^m \circ \tau_{t,s}^{(L)}(B)\}_{L \in \mathbb{R}_0^+}$, are Cauchy nets in $\mathcal{U}$ for any fixed $s, t \in \mathbb{R}$ and $m \in \mathbb{N}$. At $m = 0$, the limit is $\tau_{t,s}(B)$. As the operator δ_s is closed, by induction, for any $m \in \mathbb{N}$ and $s, t \in \mathbb{R}$, $\tau_{t,s}(B) \in \mathrm{Dom}(\delta_s^m)$ and $\delta_s^m \circ \tau_{t,s}^{(L)}(B)$ converges to $\delta_s^m \circ \tau_{t,s}(B)$, as $L \to \infty$.

(ii) For any $m \in \mathbb{N}$, $B \in \mathcal{U}_0$, and sufficiently large $L \in \mathbb{R}_0^+$,

$$\left\| \delta_s^m \circ \tau_{t,s}^{(L)}(B) \right\|_{\mathcal{U}}$$

$$\leq \sum_{\ell=0}^m \sum_{\pi \in \mathcal{S}_{\ell,m}} \sum_{x_{\pi(\ell)} \in \mathfrak{L}} \cdots \sum_{x_{\pi(m)} \in \mathfrak{L}} \sum_{\mathcal{Z}_1 \in \mathcal{P}_f(\mathfrak{L})} \cdots \sum_{\mathcal{Z}_{\pi(\ell)-1} \in \mathcal{P}_f(\mathfrak{L})} \sum_{\mathcal{Z}_{\pi(\ell)+1} \in \mathcal{P}_f(\mathfrak{L})} \cdots$$

$$\cdots \sum_{\mathcal{Z}_{\pi(m)-1} \in \mathcal{P}_f(\mathfrak{L})} \sum_{\mathcal{Z}_{\pi(m)+1} \in \mathcal{P}_f(\mathfrak{L})} \cdots \sum_{\mathcal{Z}_m \in \mathcal{P}_f(\mathfrak{L})}$$

$$\left\| \left[\Psi_{\mathcal{Z}_1}^{(s)}, \ldots, \Psi_{\mathcal{Z}_{\pi(\ell)-1}}^{(s)}, \mathbf{V}_{\{x_{\pi(\ell)}\}}^{(s)}, \Psi_{\mathcal{Z}_{\pi(\ell)+1}}^{(s)}, \right. \right.$$

$$\left. \left. \ldots, \Psi_{\mathcal{Z}_{\pi(m)-1}}^{(s)}, \mathbf{V}_{\{x_{\pi(m)}\}}^{(s)}, \Psi_{\mathcal{Z}_{\pi(m)+1}}^{(s)}, \ldots, \Psi_{\mathcal{Z}_m}^{(s)}, \tau_{t,s}^{(L)}(B) \right]^{(m+1)} \right\|_{\mathcal{U}} \,,$$

with $\mathcal{S}_{\ell,m}$ being defined by (4.49) for $\ell \in \{1, \ldots, m\}$. For $\ell = 0$, we use here the convention $\mathcal{S}_{0,m} \doteq \emptyset$ and all sums involving the maps π in the r.h.s. of the above inequality disappear in this case. Similar to (5.47), Lieb–Robinson bounds for multi-commutators imply that, if $B \in \mathcal{U}_{\Lambda_{m_0}}$, $m_0 \in \mathbb{N}_0$, then the r.h.s. of the above inequality is bounded by $D(m!)^d r^m e^{m_0 s} \|B\|_{\mathcal{U}}$, uniformly for s, t on compacta, where $r \equiv r_{d,\mathrm{T},\Psi,\mathbf{V},\mathbf{F}} \in \mathbb{R}^+$ and $D \equiv D_{\mathrm{T},\Psi,\mathbf{V}} \in \mathbb{R}^+$. We omit the details. By Assertion (i), the same bound thus holds for the norm $\|\delta_s^m \circ \tau_{t,s}(B)\|_{\mathcal{U}}$ of the limiting vector. ∎

The assumptions of Theorem 5.6 are satisfied for interactions $\Psi^{(t)} \in \mathcal{W}$ with (4.58). Note additionally that Theorem 5.6 for $s = t$ shows that

$$\mathcal{U}_0 \subseteq \bigcap_{s \in \mathbb{R}, m \in \mathbb{N}} \mathrm{Dom}\left(\delta_s^m\right) \subset \mathcal{U} \,.$$

In fact, $\mathcal{U}_0$ is a common core for $\{\delta_s\}_{s \in \mathbb{R}}$ and thus the intersection of domains

$$\bigcap_{s \in \mathbb{R}, m \in \mathbb{N}} \mathrm{Dom}\left(\delta_s^m\right) \subset \mathcal{U}$$

is also a common core of $\{\delta_s\}_{s \in \mathbb{R}}$. Observe that, at fixed $s \in \mathbb{R}$, the dense space

$$\mathrm{Dom}\left(\delta_s^\infty\right) \doteq \bigcap_{m \in \mathbb{N}} \mathrm{Dom}\left(\delta_s^m\right) \subset \mathcal{U}$$

is always a core of δ_s. See, e.g., [EN, Chap. II, 1.8 Proposition].

5.3 Application to Response Theory

In the present subsection we extend to the time-dependent case the assertions of Sect. 4.5. As previously discussed, these results can be proven, also in the non-autonomous case, for more general (time-dependent) perturbations of the form (4.70). See also proofs of Inequality (5.33) and Theorem 5.6. Similar to Sect. 4.5, the case of perturbations considered below is the relevant one to study linear and non-linear responses of interacting fermions to time-dependent external electromagnetic fields.

Let $\Psi \in \mathcal{W}$ and $\mathbf{V}$ be a potential. [So, these objects do *not* depend on time.] For any $l \in \mathbb{R}_0^+$, we consider a map $(\eta, t) \mapsto \mathbf{W}_t^{(l,\eta)}$ from $\mathbb{R}^2$ to the subspace of self-adjoint elements of $\mathcal{U}_{\Lambda_l}$. Like (4.68), we consider elements of the form

$$\mathbf{W}_t^{(l,\eta)} \doteq \sum_{x \in \Lambda_l} \sum_{z \in \mathfrak{L}, |z| \leq 1} \mathbf{w}_{x,x+z}(\eta, t) a_x^* a_{x+z} , \qquad (\eta, t) \in \mathbb{R}^2, \; l \in \mathbb{R}_0^+ , \qquad (5.48)$$

where $\{\mathbf{w}_{x,y}\}_{x,y \in \mathfrak{L}}$ are complex-valued functions of $(\eta, t) \in \mathbb{R}^2$ with

$$\overline{\mathbf{w}_{x,y}(\eta, t)} = \mathbf{w}_{y,x}(\eta, t) \qquad \text{and} \qquad \mathbf{w}_{x,y}(0, t) = 0 \qquad (5.49)$$

for all $x, y \in \mathfrak{L}$ and $(\eta, t) \in \mathbb{R}^2$. We assume that $\{\mathbf{w}_{x,y}(\eta, \cdot)\}_{x,y \in \mathfrak{L}, \eta \in \mathbb{R}}$ is a family of continuous and uniformly bounded functions (of time): There is $K_1 \in \mathbb{R}^+$ such that

$$\sup_{x,y \in \mathfrak{L}} \sup_{\eta, t \in \mathbb{R}} \left| \mathbf{w}_{x,y}(\eta, t) \right| \leq K_1 . \qquad (5.50)$$

The self-adjoint elements $\mathbf{W}_t^{(l,\eta)}$ of $\mathcal{U}$ are related to perturbations of dynamics caused by time-dependent external electromagnetic fields that vanish outside the box Λ_l. By the above conditions on $\mathbf{w}_{x,y}$, for all $l, \eta \in \mathbb{R}$, $t \mapsto \mathbf{W}_t^{(l,\eta)}$ is a continuous map from $\mathbb{R}$ to $\mathcal{B}(\mathcal{U})$.

We now denote the perturbed dynamics by the family $\{\tilde{\tau}_{t,s}^{(l,\eta)}\}_{s,t\in\mathbb{R}}$ of $*$-automorphisms generated by the symmetric derivation

$$\delta_t^{(l,\eta)} \doteq \delta + i\left[\mathbf{W}_t^{(l,\eta)}, \cdot\right] , \qquad l \in \mathbb{R}_0^+, \ \eta \in \mathbb{R} , \tag{5.51}$$

in the sense of Corollary 5.2. [This family of $*$-automorphisms has *nothing* to do with (5.10).] Recall that δ is the symmetric derivation of Theorem 4.8. The last term in the r.h.s. of (5.51) is clearly a perturbation of δ which depends continuously on time, in the sense of the operator norm on $\mathcal{B}(\mathcal{U})$. It is easy to prove in this case that $\{\tilde{\tau}_{t,s}^{(l,\eta)}\}_{s,t\in\mathbb{R}}$ is the unique *fundamental solution* of (5.18). It means that $\{\tilde{\tau}_{t,s}^{(l,\eta)}\}_{s,t\in\mathbb{R}}$ is strongly continuous, conserves the domain

$$\mathrm{Dom}(\delta_t^{(l,\eta)}) = \mathrm{Dom}(\delta) ,$$

satisfies

$$\tilde{\tau}_{t,\cdot}^{(l,\eta)}(B) \in C^1(\mathbb{R}; (\mathrm{Dom}(\delta), \|\cdot\|_{\mathcal{U}})) , \quad \tilde{\tau}_{\cdot,s}^{(l,\eta)}(B) \in C^1(\mathbb{R}; (\mathrm{Dom}(\delta), \|\cdot\|_{\mathcal{U}}))$$

for all $B \in \mathrm{Dom}(\delta)$, and solves the abstract Cauchy initial value problem (5.18) on $\mathrm{Dom}(\delta)$.

To explicitly verify this, define the family $\{\mathfrak{V}_{t,s}\}_{s,t\in\mathbb{R}} \subset \mathcal{U}$ of unitary elements by the absolutely summable series

$$\mathfrak{V}_{t,s} \doteq \mathbf{1}_{\mathcal{U}} + \sum_{k\in\mathbb{N}} i^k \int_s^t ds_1 \cdots \int_s^{s_{k-1}} ds_k \, \mathbf{W}_{s_k,s_k}^{(l,\eta)} \cdots \mathbf{W}_{s_1,s_1}^{(l,\eta)} , \tag{5.52}$$

where

$$\mathbf{W}_{t,s}^{(l,\eta)} \doteq \tau_t(\mathbf{W}_s^{(l,\eta)}) \in \mathrm{Dom}(\delta) , \qquad l \in \mathbb{R}_0^+, \ \eta, s, t \in \mathbb{R} .$$

By using this unitary family, we obtain the following additional properties of the perturbed dynamics:

Theorem 5.7 (Properties of the perturbed dynamics)
Let $\Psi \in \mathcal{W}$, $l \in \mathbb{R}_0^+$, $\eta, \eta_0 \in \mathbb{R}$, and $\mathbf{V}$ be a potential. Assume Conditions (5.49) and (5.50) with $\{\mathbf{w}_{x,y}(\eta, \cdot)\}_{x,y\in\mathfrak{L},\eta\in\mathbb{R}}$ being a family of continuous functions (of time). Then, the family $\{\tilde{\tau}_{t,s}^{(l,\eta)}\}_{s,t\in\mathbb{R}}$ of $*$-automorphisms has the following properties:
(i) Non-autonomous evolution equation. It is the unique fundamental solution of

$$\forall s, t \in \mathbb{R}: \qquad \partial_s \tilde{\tau}_{t,s}^{(l,\eta)} = -\delta_s^{(l,\eta)} \circ \tilde{\tau}_{t,s}^{(l,\eta)} , \qquad \tilde{\tau}_{t,t}^{(l,\eta)} = \mathbf{1}_{\mathcal{U}} .$$

(ii) Interaction picture. For any $s, t \in \mathbb{R}$,

$$\tilde{\tau}_{t,s}^{(l,\eta)}(B) = \tau_{-s}\left(\mathfrak{V}_{t,s}\tau_t(B)\mathfrak{V}_{t,s}^*\right) , \qquad B \in \mathcal{U} .$$

(iii) *Dyson–Phillips series. For any $s, t \in \mathbb{R}$ and $B \in \mathcal{U}$,*

$$\tilde{\tau}_{t,s}^{(l,\eta)}(B) = \tilde{\tau}_{t,s}^{(l,\eta_0)}(B) + \sum_{k \in \mathbb{N}} i^k \int_s^t ds_1 \cdots \int_s^{s_{k-1}} ds_k \tag{5.53}$$

$$\left[\mathbf{X}_{s_k,s,s_k}^{(l,\eta_0,\eta)}, \ldots, \mathbf{X}_{s_1,s,s_1}^{(l,\eta_0,\eta)}, \tilde{\tau}_{t,s}^{(l,\eta_0)}(B) \right]^{(k+1)}.$$

Here, the series absolutely converges and

$$\mathbf{X}_{t,s,\alpha}^{(l,\eta_0,\eta)} \doteq \tilde{\tau}_{t,s}^{(l,\eta_0)} \left(\mathbf{W}_\alpha^{(l,\eta)} - \mathbf{W}_\alpha^{(l,\eta_0)} \right), \quad l \in \mathbb{R}_0^+, \ \alpha, s, t, \eta_0, \eta \in \mathbb{R}. \tag{5.54}$$

Proof Before starting, note that Assertion (i) cannot be deduced from Theorem 5.5 because the cases for which (4.10) holds for some time $t \in \mathbb{R}$ is excluded by assumptions of that theorem.

1. Assertion (i) could be deduced from [K2, Theorem 6.1]. Here, we use [BB, Theorem 88] because it is proven from three conditions (B1–B3) that are elementary to verify:

B1 *(Kato quasi-stability).* For any $t \in \mathbb{R}$, the generator $\delta_t^{(l,\eta)}$ is conservative, by Lemma 4.5, and Condition B1 of [BB, Sect. 7.1] is clearly satisfied for $\lambda_1, \ldots, \lambda_n \in \mathbb{R}^+$, even with *non-ordered* and all real times $t_1, \ldots, t_n \in \mathbb{R}$. Indeed, $\{\delta_t^{(l,\eta)}\}_{t \in \mathbb{R}}$, $l \in \mathbb{R}_0^+$, generate strongly continuous groups, and not only C_0-semigroups.

B2 *(Domains and continuity).* $\{\mathbf{w}_{x,y}(\eta, \cdot)\}_{x,y \in \mathfrak{L}, \eta \in \mathbb{R}}$ is by assumption a family of continuous functions (of time) and thus, the map $t \mapsto [\mathbf{W}_t^{(l,\eta)}, \cdot]$ from $\mathbb{R}$ to $\mathcal{B}(\mathcal{U})$ is continuous in operator norm. It follows that Condition B2 of [BB, Sect. 7.1] holds with the Banach space

$$\mathcal{Y} \doteq (\mathrm{Dom}(\delta), \|\cdot\|_\delta), \tag{5.55}$$

$\|\cdot\|_\delta$ being the graph norm of the closed operator δ.

B3 *(Intertwining condition).* Since δ is a symmetric derivation with core $\mathcal{U}_0$ (Theorem 4.8 (ii)) and $\mathbf{W}_t^{(l,\eta)} \in \mathcal{U}_{\Lambda_l}$, for any $l \in \mathbb{R}_0^+$, $\eta \in \mathbb{R}$, $t \in \mathbb{R}$ and $B \in \mathrm{Dom}(\delta)$,

$$\delta \left(\left[\mathbf{W}_t^{(l,\eta)}, B \right] \right) - \left[\mathbf{W}_t^{(l,\eta)}, \delta(B) \right] = \left[\delta \left(\mathbf{W}_t^{(l,\eta)} \right), B \right] \in \mathcal{U}$$

while, by using (4.28), one verifies that

$$\left\| \left[\delta \left(\mathbf{W}_t^{(l,\eta)} \right), B \right] \right\|_{\mathcal{U}} \leq 4 \|B\|_{\mathcal{U}} \|\mathbf{W}_t^{(l,\eta)}\|_{\mathcal{U}}$$

$$\times \left(|\Lambda_l| \, \mathbf{F}(0) \, \|\Psi\|_{\mathcal{W}} + \sum_{x \in \Lambda_l} \|\mathbf{V}_{\{x\}}\|_{\mathcal{U}} \right).$$

In particular, Condition B3 of [BB, Sect. 7.1] holds true with $\Theta = \delta$.

Therefore, similar to [BB, Theorem 70 (v)], we infer from an extension of [BB, Theorem 88], which takes into account the fact that B1 holds with non-ordered real times (see, e.g., the proof of [BB, Lemma 89]), the existence of a unique solution $\{\mathfrak{W}_{s,t}\}_{s,t\in\mathbb{R}}$ of the non-autonomous evolution equation

$$\forall s,t \in \mathbb{R}: \quad \partial_s \mathfrak{W}_{s,t} = -\delta_s^{(l,\eta)} \circ \mathfrak{W}_{s,t}, \quad \mathfrak{W}_{t,t} = \mathbf{1}_{\mathcal{U}}, \tag{5.56}$$

in the strong sense on $\mathrm{Dom}(\delta) \subset \mathcal{U}$. Here, $\{\mathfrak{W}_{s,t}\}_{s,t\in\mathbb{R}}$ is an evolution family of $\mathcal{B}(\mathcal{U})$, that is, a strongly continuous two-parameter family of bounded operators acting on $\mathcal{U}$ that satisfies the cocycle (Chapman–Kolmogorov) property

$$\forall t,r,s \in \mathbb{R}: \quad \mathfrak{W}_{s,t} = \mathfrak{W}_{s,r}\mathfrak{W}_{r,t}.$$

2. Note now that the family $\{\mathfrak{V}_{t,s}\}_{s,t\in\mathbb{R}}$ was already studied in the proof of [BPH1, Theorem 5.3] for general closed symmetric derivations δ on $\mathcal{U}$: The series (5.52) absolutely converges in the Banach space $\mathcal{Y}$ (5.55). Additionally, for any $s,t \in \mathbb{R}$,

$$\partial_t \mathfrak{V}_{t,s} = i\mathfrak{V}_{t,s}\mathbf{W}_{t,t}^{(l,\eta)} \quad \text{and} \quad \partial_s \mathfrak{V}_{t,s} = -i\mathbf{W}_{s,s}^{(l,\eta)}\mathfrak{V}_{t,s}$$

hold in the sense of the Banach space $\mathcal{Y}$, and thus also in the sense of $\mathcal{U}$. Therefore, for any $s,t \in \mathbb{R}$,

$$\mathfrak{W}_{s,t}(B) = \tau_{-s}\left(\mathfrak{V}_{t,s}\tau_t(B)\mathfrak{V}_{t,s}^*\right), \quad B \in \mathcal{U}. \tag{5.57}$$

To show this equality, use the fact that the r.h.s. of this equation defines an evolution family that is a fundamental solution of (5.56), see [BPH1, Eqs. (5.24)–(5.26)].

3. Since $\{\tau_t\}_{t\in\mathbb{R}}$ is a group of $*$-automorphisms and $\{\mathfrak{V}_{t,s}\}_{s,t\in\mathbb{R}}$ is a family of unitary elements of $\mathcal{U}$, we deduce from (5.57) that $\{\mathfrak{W}_{s,t}\}_{s,t\in\mathbb{R}}$ is a collection of $*$-automorphisms of the C^*-algebra $\mathcal{U}$. We also infer from (5.57) that the two-parameter evolution family $\{\mathfrak{W}_{s,t}\}_{s,t\in\mathbb{R}}$ solves on $\mathrm{Dom}(\delta)$ the abstract Cauchy initial value problem

$$\forall s,t \in \mathbb{R}: \quad \partial_t \mathfrak{W}_{s,t} = \mathfrak{W}_{s,t} \circ \delta_t^{(l,\eta)}, \quad \mathfrak{W}_{s,s} = \mathbf{1}_{\mathcal{U}}. \tag{5.58}$$

The solution of (5.58) is unique in $\mathcal{B}(\mathcal{U})$, by Corollary 5.2 (iii). We thus arrive at Assertions (i)–(ii) with the equality

$$\tilde{\tau}_{t,s}^{(l,\eta)} = \mathfrak{W}_{s,t}, \quad l \in \mathbb{R}_0^+, \ \eta, s, t \in \mathbb{R}. \tag{5.59}$$

4. For any $l \in \mathbb{R}_0^+$, $s,t \in \mathbb{R}$, $\eta, \eta_0 \in \mathbb{R}$, and $B \in \mathcal{U}$, define

$$\hat{\tau}_{t,s}^{(l,\eta,\eta_0)} (B) \doteq \tilde{\tau}_{t,s}^{(l,\eta_0)} (B) + \sum_{k\in\mathbb{N}} i^k \int_s^t ds_1 \cdots \int_s^{s_{k-1}} ds_k \tag{5.60}$$

$$\left[\mathbf{X}_{s_k,s,s_k}^{(l,\eta_0,\eta)}, \ldots, \mathbf{X}_{s_1,s,s_1}^{(l,\eta_0,\eta)}, \tilde{\tau}_{t,s}^{(l,\eta_0)} (B) \right]^{(k+1)} .$$

This series is well-defined and absolutely convergent. Indeed, because of (5.50), there is a constant $D \in \mathbb{R}^+$ such that, for all $l \in \mathbb{R}_0^+$ and $\eta, \eta_0 \in \mathbb{R}$,

$$\sup_{t\in\mathbb{R}} \left\| \delta_t^{(l,\eta)} - \delta_t^{(l,\eta_0)} \right\|_{\mathcal{B}(\mathcal{U})} < D .$$

It follows that

$$\left\| \hat{\tau}_{t,s}^{(l,\eta,\eta_0)} \right\|_{\mathcal{B}(\mathcal{U})} \le e^{D(t-s)} , \qquad l \in \mathbb{R}_0^+, \; s,t \in \mathbb{R}, \; \eta, \eta_0 \in \mathbb{R} . \tag{5.61}$$

See, e.g., [P, Chap. 5, Theorems 2.3 and 3.1]. Now, for any $l \in \mathbb{R}_0^+$, $s, t \in \mathbb{R}$, $\eta, \eta_0 \in \mathbb{R}$, and $B \in \mathcal{U}$, note that (5.60) yields

$$\hat{\tau}_{t,s}^{(l,\eta,\eta_0)} (B) = \tilde{\tau}_{t,s}^{(l,\eta_0)} (B) + i \int_s^t ds_1 \hat{\tau}_{s_1,s}^{(l,\eta,\eta_0)} \left(\left[\mathbf{W}_{s_1}^{(l,\eta)} - \mathbf{W}_{s_1}^{(l,\eta_0)}, \tilde{\tau}_{t,s_1}^{(l,\eta_0)}(B) \right] \right)$$

from which we deduce that $\{\hat{\tau}_{t,s}^{(l,\eta)}\}_{s,t\in\mathbb{R}}$ solves (5.13), by (5.58), (5.59), (5.61) and continuity of the maps $t \mapsto \mathbf{W}_t^{(l,\eta)}$ and $t \mapsto \tilde{\tau}_{t,s}^{(l,\eta_0)}(B)$ from $\mathbb{R}$ to $\mathcal{U}$. Hence, by Corollary 5.2 (iii), $\hat{\tau}_{t,s}^{(l,\eta,\eta_0)} = \tilde{\tau}_{t,s}^{(l,\eta)}$ for any $l \in \mathbb{R}_0^+$, $s, t \in \mathbb{R}$ and $\eta, \eta_0 \in \mathbb{R}$. ∎

Now, by assuming the uniform Lipschitz continuity of the family

$$\{\mathbf{w}_{x,y}(\cdot, t)\}_{x,y\in\mathfrak{L}, t\in\mathbb{R}}$$

of functions (of η), i.e., for all parameters $\eta, \eta_0 \in \mathbb{R}$,

$$\sup_{x,y\in\mathfrak{L}} \sup_{t\in\mathbb{R}} \left| \mathbf{w}_{x,y}(\eta, t) - \mathbf{w}_{x,y}(\eta_0, t) \right| \le K_1 \left| \eta - \eta_0 \right| , \tag{5.62}$$

we can extend Theorem 4.15 to the non-autonomous case.

To this end, for some interaction Φ with energy observables $U_{\Lambda_L}^\Phi$ defined by (4.73) we study the increment (4.125), which now equals

$$\mathbf{T}_{t,s}^{(l,\eta,L)} \doteq \tilde{\tau}_{t,s}^{(l,\eta)} (U_{\Lambda_L}^\Phi) - \tau_{t,s}(U_{\Lambda_L}^\Phi) , \qquad l, L \in \mathbb{R}_0^+, \; s, t, \eta \in \mathbb{R} . \tag{5.63}$$

By (5.49), note again that $\mathbf{T}_{t,s}^{(l,0,L)} = 0$. Exactly like in the proof of Theorem 4.15, we prove a version of Taylor's theorem for increments in the non-autonomous case:

Theorem 5.8 (Taylor's theorem for increments)

Let $l, T \in \mathbb{R}_0^+$, $s, t \in [-T, T]$, $\eta, \eta_0 \in \mathbb{R}$, $\Psi \in \mathcal{W}$, and $\mathbf{V}$ be any potential. Assume (4.56) with $\varsigma > d$, (5.49)–(5.50) and (5.62), with $\{\mathbf{w}_{x,y}(\eta, \cdot)\}_{x,y \in \mathfrak{L}, \eta \in \mathbb{R}}$ being a family of continuous functions (of time). Take an interaction Φ satisfying (4.80) with $\mathbf{v}_m = (1 + m)^\varsigma$. Then:

(i) The map $\eta \mapsto \mathbf{T}_{t,s}^{(l,\eta,L)}$ converges uniformly on $\mathbb{R}$, as $L \to \infty$, to a continuous function $\mathbf{T}_{t,s}^{(l,\eta)}$ of η and

$$\mathbf{T}_{t,s}^{(l,\eta)} - \mathbf{T}_{t,s}^{(l,\eta_0)} = \sum_{\Lambda \in \mathcal{P}_f(\mathfrak{L})} i \int_s^t ds_1 \tilde{\tau}_{s_1,s}^{(l,\eta)} \left(\left[\mathbf{W}_{s_1}^{(l,\eta)} - \mathbf{W}_{s_1}^{(l,\eta_0)}, \tilde{\tau}_{t,s_1}^{(l,\eta_0)}(\Phi_\Lambda) \right] \right) .$$

(ii) For any $m \in \mathbb{N}$ satisfying $d(m + 1) < \varsigma$,

$$\mathbf{T}_{t,s}^{(l,\eta)} - \mathbf{T}_{t,s}^{(l,\eta_0)}$$

$$= \sum_{k=1}^m \sum_{\Lambda \in \mathcal{P}_f(\mathfrak{L})} i^k \int_s^t ds_1 \cdots \int_s^{s_{k-1}} ds_k \left[\mathbf{X}_{s_k,s,s_k}^{(l,\eta_0,\eta)}, \ldots, \mathbf{X}_{s_1,s,s_1}^{(l,\eta_0,\eta)}, \tilde{\tau}_{t,s}^{(l,\eta_0)}(\Phi_\Lambda) \right]^{(k+1)}$$

$$+ \sum_{\Lambda \in \mathcal{P}_f(\mathfrak{L})} i^{m+1} \int_s^t ds_1 \cdots \int_s^{s_m} ds_{m+1}$$

$$\tilde{\tau}_{s_{m+1},s}^{(l,\eta)} \left(\left[\mathbf{W}_{s_{m+1}}^{(l,\eta)} - \mathbf{W}_{s_{m+1}}^{(l,\eta_0)}, \mathbf{X}_{s_m,s_{m+1},s_m}^{(l,\eta_0,\eta)}, \ldots, \mathbf{X}_{s_1,s_{m+1},s_1}^{(l,\eta_0,\eta)}, \tilde{\tau}_{t,s_{m+1}}^{(l,\eta_0)}(\Phi_\Lambda) \right]^{(m+2)} \right) .$$

(iii) All the above series in Λ absolutely converge: For any $m \in \mathbb{N}$ satisfying $d(m + 1) < \varsigma$, $k \in \{1, \ldots, m\}$, and $\{s_j\}_{j=1}^{m+1} \subset [-T, T]$,

$$\sum_{\Lambda \in \mathcal{P}_f(\mathfrak{L})} \left\| \left[\mathbf{X}_{s_k,s,s_k}^{(l,\eta_0,\eta)}, \ldots, \mathbf{X}_{s_1,s,s_1}^{(l,\eta_0,\eta)}, \tilde{\tau}_{t,s}^{(l,\eta_0)}(\Phi_\Lambda) \right]^{(k+1)} \right\|_\mathcal{U} \leq D |\Lambda_l| |\eta - \eta_0|^k$$

and

$$\sum_{\Lambda \in \mathcal{P}_f(\mathfrak{L})} \left\| \tilde{\tau}_{s_{m+1},s}^{(l,\eta)} \left(\left[\mathbf{W}_{s_{m+1}}^{(l,\eta)} - \mathbf{W}_{s_{m+1}}^{(l,\eta_0)}, \mathbf{X}_{s_m,s_{m+1},s_m}^{(l,\eta_0,\eta)}, \ldots, \mathbf{X}_{s_1,s_{m+1},s_1}^{(l,\eta_0,\eta)}, \tilde{\tau}_{t,s_{m+1}}^{(l,\eta_0)}(\Phi_\Lambda) \right]^{(m+2)} \right) \right\|_\mathcal{U}$$

$$\leq D |\Lambda_l| |\eta - \eta_0|^{m+1}$$

for some constant $D \in \mathbb{R}^+$ depending only on $m, d, T, \Psi, K_1, \Phi, \mathbf{F}$. The last assertion also holds for $m = 0$.

Proof By Theorems 4.11 and 5.4, Corollary 4.12 holds in the non-autonomous case. Moreover, by Lemmas 5.3 and 4.9 is also satisfied in the non-autonomous case. Therefore, the proof is an easy extension of the proof of Theorem 4.15. ∎

If the interaction has exponential decay, we show that the map $\eta \mapsto |\Lambda_l|^{-1} \mathbf{T}_{t,s}^{(l,\eta)}$ from $\mathbb{R}$ to $\mathcal{U}$ is bounded in the sense of Gevrey classes, uniformly w.r.t. $l \in \mathbb{R}_0^+$. This corresponds to Theorem 4.16 in the non-autonomous case:

Theorem 5.9 (Increments as Gevrey maps)
Let $l, \mathrm{T} \in \mathbb{R}_0^+$, $s, t \in [-\mathrm{T}, \mathrm{T}]$, $\Psi \in \mathcal{W}$, *and* $\mathbf{V}$ *be any potential. Assume (4.57) and take an interaction* Φ *satisfying (4.80) with* $\mathbf{v}_m = \mathrm{e}^{m\varsigma}$. *For all* $x, y \in \mathfrak{L}$, *assume further the real analyticity of the map* $\eta \mapsto \mathbf{w}_{x,y}(\eta, \cdot)$ *from* $\mathbb{R}$ *to the Banach space* $C(\mathbb{R}; \mathbb{C})$, *which is equipped with the supremum norm, as well as the existence of* $r \in \mathbb{R}^+$ *such that*

$$K_2 \doteq \sup_{x,y \in \mathfrak{L}} \sup_{m \in \mathbb{N}} \sup_{\eta, t \in \mathbb{R}} \frac{r^m \partial_\eta^m \mathbf{w}_{x,y}(\eta, t)}{m!} < \infty .$$

(i) Smoothness. As a function of $\eta \in \mathbb{R}$, $\mathbf{T}_{t,s}^{(l,\eta)} \in C^\infty(\mathbb{R}; \mathcal{U})$ *and for any* $m \in \mathbb{N}$,

$$\partial_\eta^m \mathbf{T}_{t,s}^{(l,\eta)} = \sum_{k=1}^m \sum_{\Lambda \in \mathcal{P}_f(\mathfrak{L})} i^k \int_s^t \mathrm{d}s_1 \cdots \int_s^{s_{k-1}} \mathrm{d}s_k$$
$$\partial_\varepsilon^m \left[\mathbf{X}_{s_k,s,s_k}^{(l,\eta,\eta+\varepsilon)}, \ldots, \mathbf{X}_{s_1,s,s_1}^{(l,\eta,\eta+\varepsilon)}, \tilde{\tau}_{t,s}^{(l,\eta)}(\Phi_\Lambda) \right]^{(k+1)} \bigg|_{\varepsilon=0} .$$

The above series in Λ *are absolutely convergent.*
(ii) Uniform boundedness of the Gevrey norm of density of increments. There exist $\tilde{r} \equiv \tilde{r}_{d,\mathrm{T},\Psi,K_2,\mathrm{F}} \in \mathbb{R}^+$ *and* $D \equiv D_{\mathrm{T},\Psi,K_2,\Phi} \in \mathbb{R}^+$ *such that, for all* $l \in \mathbb{R}_0^+$, $\eta \in \mathbb{R}$ *and* $s, t \in [-\mathrm{T}, \mathrm{T}]$,

$$\sum_{m \in \mathbb{N}} \frac{\tilde{r}^m}{(m!)^d} \sup_{l \in \mathbb{R}_0^+} \left\| |\Lambda_l|^{-1} \partial_\eta^m \mathbf{T}_{t,s}^{(l,\eta)} \right\|_\mathcal{U} \le D .$$

Proof Like for Theorem 5.8, the assertions are easily proven by extending the proof of Theorem 4.16 to the non-autonomous case. ∎

This theorem has important consequences in terms of increment density limit

$$\lim_{l \to \infty} |\Lambda_l|^{-1} \rho(\mathbf{T}_{t,s}^{(l,\eta)})$$

at any fixed $s, t \in \mathbb{R}$ and state $\rho \in \mathcal{U}^*$. This limit is to be understood as an accumulation point of the bounded net $\{|\Lambda_l|^{-1} \rho(\mathbf{T}_{t,s}^{(l,\eta)})\}_{l>0}$:

Corollary 5.10 (Increment density limit)
Let $\rho \in \mathcal{U}^*$. *Under the conditions of Theorem 5.9, there is a subsequence* $\{l_n\}_{n \in \mathbb{N}} \subset \mathbb{R}_0^+$ *such that, for all* $s, t \in [-\mathrm{T}, \mathrm{T}]$, *the following limit exists*

$$\eta \mapsto \mathbf{g}_{t,s}(\eta) \doteq \lim_{n \to \infty} |\Lambda_{l_n}|^{-1} \rho(\mathbf{T}_{t,s}^{(l_n,\eta)})$$

and defines a smooth function $\mathbf{g}_{t,s} \in C^\infty(\mathbb{R})$. Furthermore, there exist $\tilde{r} \equiv \tilde{r}_{d,T,\Psi,K_2,F} \in \mathbb{R}^+$ and $D \equiv D_{T,\Psi,K_2,\Phi} \in \mathbb{R}^+$ such that, for all $\eta \in \mathbb{R}$ and $s, t \in [-T, T]$,

$$\sum_{m \in \mathbb{N}} \frac{\tilde{r}^m}{(m!)^d} \left| \partial_\eta^m \mathbf{g}_{t,s}(\eta) \right| \leq D .$$

Proof Let $T \in \mathbb{R}_0^+$. By Theorem 5.8 (i) for $\eta_0 = 0$ together with (5.49) and Corollary 5.2 (ii),

$$\sup_{l \in \mathbb{R}_0^+} \sup_{\eta \in \mathbb{R}} \sup_{s,t \in [-T,T]} \left\{ |\Lambda_l|^{-1} \rho(\mathbf{T}_{t,s}^{(l,\eta)}) \right\} < \infty . \tag{5.64}$$

Furthermore, we infer from Theorem 5.9 that, for any $m \in \mathbb{N}$,

$$\sup_{l \in \mathbb{R}_0^+} \sup_{\eta \in \mathbb{R}} \sup_{s,t \in [-T,T]} \left\{ |\Lambda_l|^{-1} \rho(\partial_\eta^m \mathbf{T}_{t,s}^{(l,\eta)}) \right\} < \infty . \tag{5.65}$$

By (5.64) and (5.65), the assertions are consequences of Theorem 5.9 combined with the mean value theorem and the (Arzelà–) Ascoli theorem [Ru, Theorem A5]. Indeed, $\{l_n\}_{n \in \mathbb{N}} \subset \mathbb{R}_0^+$ is taken as a so-called diagonal sequence $l_n = l_n^{(n)}$ of a family $\{l_n^{(m)}\}_{n \in \mathbb{N}}$, $m \in \mathbb{N}_0$, of sequences in $\mathbb{R}_0^+$ such that, for all $m \in \mathbb{N}_0$, the m–th derivative $|\Lambda_{l_n}|^{-1} \partial_\eta^m \mathbf{T}_{t,s}^{(l_n^{(m)},\eta)}$ uniformly converges as $n \to \infty$. With this choice,

$$\partial_\eta^m \mathbf{g}_{t,s}(\eta) = \lim_{n \to \infty} |\Lambda_{l_n}|^{-1} \rho(\partial_\eta^m \mathbf{T}_{t,s}^{(l_n,\eta)}) .$$

$\blacksquare$

From the above corollary, at dimension $d = 1$ and for s, t on compacta, the increment density limit $\mathbf{g}_{t,s} \in C^\infty(\mathbb{R})$ defines a real analytic function. As a consequence, the increment density limit is never zero for η outside a discrete subset of $\mathbb{R}$, unless $\mathbf{g}_{t,s}$ is *identically* vanishing for *all* $\eta \in \mathbb{R}$.

This mathematical property refers to a physical one. It reflects a generic alternative between either *strictly* positive or *identically* vanishing heat production density, at macroscopic scale, in presence of non-vanishing external electric fields. Indeed, by taking $\Phi = \Psi$ in Theorem 5.9, $\mathbf{T}_{t,s}^{(l,\eta)}$ is related to the heat produced by the presence of an electromagnetic field, encoded in $\mathbf{W}_t^{(l,\eta)}$. If we use cyclic processes, which means here that $\mathbf{W}_t^{(l,\eta)} = 0$ outside some compact set $[t_0, t_1] \subset \mathbb{R}$, then the KMS state $\varrho \in \mathcal{U}^*$ applied on the energy increment $\mathbf{T}_{t_1,t_0}^{(l,\eta)}$ is the total heat production (1st law of Thermodynamics) with increment density limit equal to $\mathbf{g}_{t_1,t_0}(\eta)$. It is non-negative, by the 2nd law of Thermodynamics. See [BP5] for more details on the 1st and 2nd laws for the quantum systems considered here. Now, if $\mathbf{g}_{t_1,t_0}(\eta)$ is *identically* vanishing for *all* $\eta \in \mathbb{R}$ then it means that the external perturbation never produces heat in the system, which is a very strong property. The latter is expected to be the case, for instance, for superconductors driven by electric perturbations. This kind of behavior should highlight major features of the system (like possibly broken symmetry). Hence, if the heat production density is not *identically* vanishing, generically, it is *strictly* positive, at least at dimension $d = 1$, because of properties of real analytic functions mentioned above.

For higher dimensions $d > 1$ and s, t on compacta, Corollary 5.10 implies that the increment density limit $\mathbf{g}_{t,s} \in C^\infty(\mathbb{R})$ belongs to the Gevrey class

$$C_d^\omega(\mathbb{R}) \doteq \left\{ f \in C^\infty(\mathbb{R}) \ : \ \sup_{\eta \in \mathbb{R}} \left| \partial_\eta^m f(\eta) \right| \leq D^m (m!)^d \text{ for any } m \in \mathbb{N} \right\}.$$

If $d > 1$, the elements of $C_d^\omega(\mathbb{R})$ are usually neither analytic nor quasi-analytic. In particular, functions of $C_d^\omega(\mathbb{R})$ can have arbitrarily small support, while $C_d^\omega(\mathbb{R}) \subsetneq C_{d'}^\omega(\mathbb{R})$ whenever $d < d'$. Thus, the alternative above, which is related to the heat production density in presence of external electric fields, does not follow from Corollary 5.10 for higher dimensions $d \geq 2$. However, note that, at least for the quasi-free dynamics (also in the presence of a random potential), the heat production density is a real analytic function of η at any dimension $d \in \mathbb{N}$, at least for η near zero. This follows from [BPH1, Theorem 3.4]. Therefore, the above alternative for the heat production density may be true at any dimension, provided the interaction decays fast enough in space (or is finite-range, in the extreme case).

Observe finally that if a Gevrey function $f : \mathbb{R} \to \mathbb{R}$ is invertible on some open interval $I \subset \mathbb{R}$ then the inverse $f^{-1} : f(I) \to \mathbb{R}$ is again a Gevrey function. So, the above theorem implies that, if the relation between applied field strength η and the density of increment at $l \to \infty$ is injective for some range of field strengths η, then the applied field strength in that range is a Gevrey function of the density of increment. For more details on Gevrey classes, see, e.g., [H].

Chapter 6
Applications to Conductivity Measures

6.1 Charged Transport Properties in Mathematics

Altogether, the classical theory of linear conductivity (including the theory of (Landau) Fermi liquids, see, e.g., [BP4] for a historical perspective) is more like a makeshift theoretical construction than a smooth and complete theory. It is unsatisfactory to use the Drude (or the Drude–Lorentz) model – which does not take into account quantum mechanics – together with certain ad hoc hypotheses as a proper microscopic explanation of conductivity. For instance, in [NS1, NS2, SE, YRMK], the (normally fixed) relaxation time of the Drude model has to be taken as an effective frequency–dependent parameter to fit with experimental data [T] on usual metals like gold. In fact, as claimed in the famous paper [So, p. 505], "*it must be admitted that there is no entirely rigorous quantum theory of conductivity.*"

Concerning AC–conductivity, however, in the last years significant mathematical progress has been made. See, e.g., [KLM, KM1, KM2, BC, BPH1, BPH2, BPH3, BPH4, BP5, BP6, W, DG] for examples of mathematically rigorous derivations of linear conductivity from first principles of quantum mechanics in the AC–regime. In particular, the notion of conductivity measure has been introduced for the first time in [KLM], albeit only for non–interacting systems. These results indicate a physical picture of the microscopic origin of Ohm and Joule's laws which differs from usual explanations coming from the Drude (Lorentz–Sommerfeld) model.

As electrical resistance of conductors may result from the presence of interactions between charge carriers, an important issue is to tackle the interacting case. This is first[1] done in [BP5, BP6] for very general systems of interacting quantum particles on lattices, including many important models of condensed matter physics like the celebrated Hubbard model. This was out of scope of [KLM, KM1, KM2, DG, BPH1,

---

[1] With regard to interacting systems, explicit constructions of KMS states are obtained in the Ph.D. thesis [W] for a one–dimensional model of interacting fermions with a finite range pair interaction. But, the author studies in [W, Chap. 9] the linear response theory only for non–interacting fermions, keeping in mind possible generalizations to interacting systems.

© The Author(s) 2017
J.-B. Bru and W. de Siqueira Pedra, *Lieb-Robinson Bounds for Multi-commutators and Applications to Response Theory*, SpringerBriefs in Mathematical Physics, DOI 10.1007/978-3-319-45784-0_6

BPH2, BPH3, BPH4, W] which strongly rely on properties of quasi–free dynamics and states.

The central issue in [BP5, BP6] is to get estimates on transport coefficients related to electric conduction, which are *uniform* w.r.t. the random parameters and the volume $|\Lambda_l|$ of the box Λ_l where the electromagnetic field lives. This is crucial to get valuable information on conductivity in the macroscopic limit $l \to \infty$ and otherwise the results presented in [BP5, BP6] would loose almost all their interest. To get such estimates in the non–interacting case [BPH1, BPH2, BPH3, BPH4], we applied tree–decay bounds on multi–commutators in the sense of [BPH1, Sect. 4]. The latter are based on combinatorial results [BPH1, Theorem 4.1] already used before, for instance in [FMU], and require the dynamics to be implemented by Bogoliubov automorphisms. A solution to the issue for the *interacting* case is made possible by the results of Sects. 4.5 and 5.3, which are direct consequences of the Lieb–Robinson bounds for multi–commutators. Detailed discussions on the estimates for the interacting case are found in [BP5]. See also Corollary 4.12, which is an extension of the tree–decay bounds [BPH1, Sect. 4] to the interacting case.

In [BP6] the existence of macroscopic AC–conductivity measures for interacting systems is derived from the 2nd law of thermodynamics, explained in Sect. 6.4. The Lieb–Robinson bound for multi–commutators of order 3 implies that it is always a Lévy measure, see [BP6, Theorems 7.1 and 5.2]. We also derive below other properties of the AC–conductivity measures from Lieb–Robinson bounds for multi–commutators of higher orders. See Sects. 6.5–6.6. In particular, we study their behavior at high frequencies (Theorems 6.1 and 6.5): in contrast to the prediction of the Drude (Lorentz–Sommerfeld) model, widely used in physics [So, LTW] to describe the phenomenon of electrical conductivity, the conductivity measure stemming from short–range interparticle interactions has to decay rapidly at high frequencies.

The proposed mathematical approach to the problem of deriving macroscopic conductivity properties from the microscopic quantum dynamics of an infinite system of particles also yield new physical insight, beyond classical theories of conduction: a notion of current viscosity related to the interplay of paramagnetic and diamagnetic currents, heat/entropy production via different types of energy and current increments, existence of (AC–) conductivity measures from the 2nd law and (possibly) as a spectral (excitation) measure from current fluctuations are all examples of new physical concepts derived in the course of the studies performed in [BPH2, BPH4, BP5, BP6] and previously not discussed in the literature.

Note, however, that, by now, our results do not give explicit information on the conductivity measure for concrete models (like the Hubbard model, for instance). The latter belongs to "hard analysis", by contrast with our results which are rather on the side of the "soft analysis" (similar to the difference between knowing the spectrum of a concrete self–adjoint operator and knowing the spectral theorem). Moreover, our approach does not directly provide a mathematical understanding from first principles of Ohm's laws as a bulk property in the DC–regime, which is one of the most important and difficult problems in mathematical physics for more than one century. We believe, however, that our results can support further rigorous developments towards a solution of such a difficult problem: one could, for instance, try to show, for some class of models, that the conductivity measure is absolutely continuous w.r.t. to the Lebesgue measure and that its Radon–Nikodym derivative is continuous at low frequencies, having a well-defined zero–frequency limit.

We thus present in the following some central results of [BP5, BP6], with a few complementary studies, as an example of an important application in mathematical physics of Lieb–Robinson bounds for multi–commutators.

6.2 Interacting Fermions in Disordered Media

(i) Kinetic part: Let $\Delta_d \in \mathcal{B}(\ell^2(\mathfrak{L}))$ be (up to a minus sign) the usual d–dimensional discrete Laplacian defined by

$$[\Delta_d(\psi)](x) \doteq 2d\psi(x) - \sum_{z \in \mathfrak{L},\, |z|=1} \psi(x+z) , \qquad x \in \mathfrak{L},\ \psi \in \ell^2(\mathfrak{L}) .$$

To understand how such terms come about by starting from the usual Laplacian in the continuum, see for instance [Ne, Sect. 2 B] which derives effective models on lattices by using so–called Wannier functions in a band subspace. This defines a short–range interaction $\Psi^{(d)} \in \mathcal{W}$ by

$$\Psi_\Lambda^{(d)} \doteq \langle \mathfrak{e}_x, \Delta_d \mathfrak{e}_y \rangle_{\ell^2(\mathfrak{L})} a_x^* a_y + \left(1 - \delta_{x,y}\right) \langle \mathfrak{e}_y, \Delta_d \mathfrak{e}_x \rangle_{\ell^2(\mathfrak{L})} a_y^* a_x \in \mathcal{U}^+ \cap \mathcal{U}_\Lambda$$

whenever $\Lambda = \{x, y\}$ for $x, y \in \mathfrak{L}$, and $\Psi_\Lambda^{(d)} \doteq 0$ otherwise. Recall that $\{\mathfrak{e}_x\}_{x \in \Lambda}$ is the (canonical) orthonormal basis of $\ell^2(\Lambda)$ defined by (3.2).

(ii) Disordered media: Disorder in the crystal is modeled by a random potential associated with a probability space $(\Omega, \mathfrak{A}_\Omega, \mathfrak{a}_\Omega)$ defined as follows: Let $\Omega \doteq [-1, 1]^{\mathfrak{L}}$. I.e., any element of Ω is a function on lattice sites with values in $[-1, 1]$. For $x \in \mathfrak{L}$, let Ω_x be an arbitrary element of the Borel σ–algebra of the interval $[-1, 1]$ w.r.t. the usual metric topology. $\mathfrak{A}_\Omega$ is the σ–algebra generated by the cylinder sets $\prod_{x \in \mathfrak{L}} \Omega_x$, where $\Omega_x = [-1, 1]$ for all but finitely many $x \in \mathfrak{L}$. Then, $\mathfrak{a}_\Omega$ is an arbitrary *ergodic* probability measure on the measurable space $(\Omega, \mathfrak{A}_\Omega)$. This means that the probability measure $\mathfrak{a}_\Omega$ is invariant under the action

$$\omega(y) \mapsto \chi_x^{(\Omega)}(\omega)(y) \doteq \omega(y + x) , \qquad x, y \in \mathbb{Z}^d , \tag{6.1}$$

of the group $(\mathbb{Z}^d, +)$ of lattice translations on Ω and, for any $\mathcal{X} \in \mathfrak{A}_\Omega$ such that $\chi_x^{(\Omega)}(\mathcal{X}) = \mathcal{X}$ for all $x \in \mathbb{Z}^d$, one has $\mathfrak{a}_\Omega(\mathcal{X}) \in \{0, 1\}$. We denote by $\mathbb{E}[\,\cdot\,]$ the expectation value associated with $\mathfrak{a}_\Omega$.

Then, any realization $\omega \in \Omega$ and strength $\lambda \in \mathbb{R}_0^+$ of disorder is implemented by the potential $\mathbf{V}^{(\omega)}$ defined by

$$\mathbf{V}_{\{x\}}^{(\omega)} \doteq \lambda \omega(x) a_x^* a_x , \qquad x \in \mathfrak{L} . \tag{6.2}$$

(iii) Interparticle interactions: They are taken into account by choosing some short–range interaction $\Psi^{IP} \in \mathcal{W}$ such that $\Psi^{IP}_\Lambda = 0$ whenever $\Lambda = \{x, y\}$ for $x, y \in \mathfrak{L}$, and

$$\sum_{\Lambda \in \mathcal{P}_f(\mathfrak{L})} [\Psi^{IP}_\Lambda, a^*_x a_x] = 0 , \qquad \Psi^{IP}_{\Lambda+x} = \chi_x (\Psi^{IP}_\Lambda) , \qquad \Lambda \in \mathcal{P}_f(\mathfrak{L}), \ x \in \mathfrak{L} . \quad (6.3)$$

Here, the family $\{\chi_x\}_{x \in \mathfrak{L}}$ of *–automorphisms of $\mathcal{U}$ implements the action of the group $(\mathbb{Z}^d, +)$ of lattice translations on the CAR C^*–algebra $\mathcal{U}$, see (4.39). Observe that this class of interparticle interactions includes all density–density interactions resulting from the second quantization of two–body interactions defined via a real–valued and summable function $v : [0, \infty) \to \mathbb{R}$ satisfying (4.9).
Then, by (i)–(iii), the full interaction

$$\Psi = \Psi^{(d)} + \Psi^{IP} \in \mathcal{W} \qquad\qquad (6.4)$$

and the potential $\mathbf{V}^{(\omega)}$ uniquely define an infinite-volume dynamics corresponding to the C_0–group $\tau^{(\omega)} \doteq \{\tau^{(\omega)}_t\}_{t \in \mathbb{R}}$ of *–automorphisms with generator $\delta^{(\omega)}$. See Theorem 4.8.

(iv) Space–homogeneous electromagnetic fields: Let $l \in \mathbb{R}^+$, $\eta \in \mathbb{R}$, and the compactly supported function $\mathcal{A} \in C^\infty_0(\mathbb{R}; \mathbb{R}^d)$ with $\mathcal{A}(t) \doteq 0$ for all $t \le 0$. Set $E(t) \doteq -\partial_t \mathcal{A}(t)$ for all $t \in \mathbb{R}$. Then, the electric field at time $t \in \mathbb{R}$ equals $\eta E(t)$ inside the cubic box Λ_l and $(0, 0, \ldots, 0)$ outside. Up to negligible terms of order $\mathcal{O}(l^{d-1})$, this leads to a perturbation (of the generator of dynamics) of the form (5.48), (5.51) with complex–valued $\{\mathbf{w}_{x,y}\}_{x,y \in \mathfrak{L}}$ functions of $(\eta, t) \in \mathbb{R}^2$ defined by $\mathbf{w}_{x,x+z}(\eta, t) = 0$ for any $x, z \in \mathfrak{L}$ with $|z| > 1$ while

$$\mathbf{w}_{x,x \pm e_q}(\eta, t) \doteq \left(\exp\left(\mp i\eta \int_0^t E_q(s) \ ds \right) - 1 \right) \langle e_x, \Delta_d e_{x \pm e_q} \rangle_{\ell^2(\mathfrak{L})} = \overline{\mathbf{w}_{x \pm e_q, x}(\eta, t)}$$

for any $q \in \{1, \ldots, d\}$. Here, $E(t) = (E_1(t), \ldots, E_d(t))$ and $\{e_q\}^d_{q=1}$ is the canonical orthonormal basis of the Euclidean space $\mathbb{R}^d$. These functions clearly satisfy Conditions (5.49)–(5.50) and (5.62). Note that such terms can be derived from the usual magnetic Laplacian (minimal coupling) in the continuum, as explained in [Ne, Sect. 2I, in particular Corollary 3.1].

Thus, the system of fermions in disordered medium, the interaction of which is encoded by (6.4), is perturbed from $t = 0$ onwards by space–homogeneous electromagnetic fields, leading to a well–defined family $\{\tilde{\tau}^{(\omega,l,\eta)}_{t,s}\}_{s,t \in \mathbb{R}}$ of *–automorphisms, as explained in Theorem 5.7.

6.3 Paramagnetic Conductivity

(i) Paramagnetic currents: For any pair $(x, y) \in \mathfrak{L}^2$, we define the current observable by

$$I_{(x,y)} \doteq i(a_y^* a_x - a_x^* a_y) = I_{(x,y)}^* \in \mathcal{U}_0 . \tag{6.5}$$

It is seen as a current because it satisfies a discrete continuity equation. See, e.g., [BP5, Sect. 3.2]. For any $\mathcal{A} \in C_0^\infty(\mathbb{R}; \mathbb{R}^d)$, $l \in \mathbb{R}^+$, $\omega \in \Omega$, $\eta \in \mathbb{R}$ and $t \in \mathbb{R}_0^+$, these observables are used to define a paramagnetic current increment density observable $\mathbb{J}_{p,l}^{(\omega)}(t, \eta) \in \mathcal{U}^d$:

$$\left\{ \mathbb{J}_{p,l}^{(\omega)}(t, \eta) \right\}_k \doteq |\Lambda_l|^{-1} \sum_{x \in \Lambda_l} \left\{ \tilde{\tau}_{t,0}^{(\omega, l, \eta)} \left(I_{(x+e_k, x)} \right) - \tau_t^{(\omega)} \left(I_{(x+e_k, x)} \right) \right\} .$$

Compare with Eq. (5.63).

Note that electric fields accelerate charged particles and induce so–called diamagnetic currents, which correspond to the ballistic movement of particles. In fact, as explained in [BPH2, Sects. III and IV], this component of the total current creates a kind of "wave front" that destabilizes the whole system by changing its state. The presence of diamagnetic currents leads then to the progressive appearance of paramagnetic currents which are responsible for heat production and the in–phase AC–conductivity of the system. Diamagnetic currents are not relevant for the present purpose and are thus not defined here. For more details, see [BPH2, BP5, BP6].

(ii) Paramagnetic conductivity: We define the space–averaged paramagnetic transport coefficient observable $\mathcal{C}_{p,l}^{(\omega)} \in C^1(\mathbb{R}; \mathcal{B}(\mathbb{R}^d; \mathcal{U}^d))$, w.r.t. the canonical orthonormal basis $\{e_q\}_{q=1}^d$ of the Euclidian space $\mathbb{R}^d$, by the corresponding matrix entries

$$\left\{ \mathcal{C}_{p,l}^{(\omega)}(t) \right\}_{k,q} \doteq \frac{1}{|\Lambda_l|} \sum_{x,y \in \Lambda_l} \int_0^t i[\tau_{-s}^{(\omega)}(I_{(y+e_q, y)}), I_{(x+e_k, x)}] ds \tag{6.6}$$

for any $l \in \mathbb{R}^+$, $\omega \in \Omega$, $t \in \mathbb{R}$ and $k, q \in \{1, \ldots, d\}$.

By (i)–(ii), if Ψ^{IP} satisfies (4.56) with $\varsigma > 2d$ (polynomial decay) then we infer from Theorem 5.8 that, for any $\mathcal{A} \in C_0^\infty(\mathbb{R}; \mathbb{R}^d)$,

$$\mathbb{J}_{p,l}^{(\omega)}(t, \eta) = \eta \mathbf{J}_{p,l}^{(\omega)}(t) + \mathcal{O}\left(\eta^2\right) . \tag{6.7}$$

The correction terms of order $\mathcal{O}(\eta^2)$ are uniformly bounded in $l \in \mathbb{R}^+$, $\omega \in \Omega$ and $\lambda, t \in \mathbb{R}_0^+$. By explicit computations, one checks that

$$\mathbf{J}_{p,l}^{(\omega)}(t) = \int_0^t \tau_t^{(\omega)} \left(\mathcal{C}_{p,l}^{(\omega)}(t - s) \right) E(s) ds \tag{6.8}$$

for any $\mathcal{A} \in C_0^\infty(\mathbb{R}; \mathbb{R}^d), l \in \mathbb{R}^+, \omega \in \Omega$ and $t \in \mathbb{R}_0^+$. The latter is the paramagnetic *linear response current*. For more details, see also [BP5, Theorem 3.7]. Here, for any $\mathbf{D} \in \mathcal{B}(\mathbb{R}^d; \mathcal{U}^d), \tau_t^{(\omega)}(\mathbf{D}) \in \mathcal{B}(\mathbb{R}^d; \mathcal{U}^d)$ is, by definition, the linear operator on $\mathbb{R}^d$ defined, w.r.t. the canonical orthonormal basis $\{e_q\}_{q=1}^d$ of the Euclidian space $\mathbb{R}^d$, by the matrix entries

$$\left\{\tau_t^{(\omega)}(\mathbf{D})\right\}_{k,q} \doteq \tau_t^{(\omega)}\left(\{\mathbf{D}\}_{k,q}\right), \qquad k, q \in \{1, \ldots, d\}.$$

6.4 2nd Law of Thermodynamics and Equilibrium States

(i) States: $\rho \in \mathcal{U}^*$ is a state if $\rho \geq 0$, that is, $\rho(B^*B) \geq 0$ for all $B \in \mathcal{U}$, and $\rho(1) = 1$. States encode the statistical distribution of all physical quantities associated with observables $B = B^* \in \mathcal{U}$. See Sect. 2.5.

For any $\mathbf{D} \in \mathcal{B}(\mathbb{R}^d; \mathcal{U}^d), \rho(\mathbf{D}) \in \mathcal{B}(\mathbb{R}^d)$ is, by definition, the linear operator defined, w.r.t. the canonical orthonormal basis $\{e_q\}_{q=1}^d$ of $\mathbb{R}^d$, by

$$\{\rho(\mathbf{D})\}_{k,q} \doteq \rho\left(\{\mathbf{D}\}_{k,q}\right), \qquad k, q \in \{1, \ldots, d\}.$$

(ii) 2nd law of thermodynamics: As explained in [LY1, LY2], different formulations of the same principle have been stated by Clausius, Kelvin (and Planck), and Carathéodory. Our study is based on the Kelvin–Planck statement while avoiding the concept of "cooling" [LY1, p. 49]. It can be expressed as follows [PW, p. 276]:
Systems in the equilibrium are unable to perform mechanical work in cyclic processes.
(iii) Passive states: To define equilibrium states, the 2nd law, as expressed in [PW], is pivotal because it leads to a clear mathematical formulation of the Kelvin–Planck notion of equilibrium: For any strongly continuous one–parameter group $\tau \equiv \{\tau_t\}_{t \in \mathbb{R}}$ of ∗–automorphisms of $\mathcal{U}$, one obtains a well–defined strongly continuous two–parameter family $\{\tau_{t,t_0}^{(\mathbf{W})}\}_{t \geq t_0}$ of ∗–automorphisms of $\mathcal{U}$ by perturbing the generator of dynamics with bounded time–dependent symmetric derivations

$$B \mapsto i\,[\mathbf{W}_t, B], \qquad B \in \mathcal{U}, t \in \mathbb{R},$$

for any arbitrary *cyclic process* $\{\mathbf{W}_t\}_{t \geq t_0}$ of time length $T \geq 0$, that is, a differentiable family $\{\mathbf{W}_t\}_{t \geq t_0} \subset \mathcal{U}$ of self–adjoint elements of $\mathcal{X}$ such that $\mathbf{W}_t = 0$ for all real times $t \notin [t_0, T + t_0]$. Then, a state $\varrho \in \mathcal{U}^*$ is *passive* (cf. [PW, Definition 1.1]) iff the work

$$\int_{t_0}^t \varrho \circ \tau_{t,t_0}^{(\mathbf{W})}(\partial_t \mathbf{W}_t)\,dt$$

performed on the system is non–negative for all cyclic processes $\{\mathbf{W}_t\}_{t \geq t_0}$ of any time length $T \geq 0$. By [PW, Theorem 1.1], such states are invariant w.r.t. the unperturbed dynamics: $\varrho = \varrho \circ \tau_t$ for any $t \in \mathbb{R}$.

If $\tau = \tau^{(\omega)}$ with $\omega \in \Omega$ then, as explained in [BP5, Sect. 2.6], at least one passive state $\varrho^{(\omega)}$ exists. It represents an equilibrium state of the system (in a broad sense), the mathematical definition of which encodes the 2nd law.

(iv) Random invariant passive states: We impose two natural conditions on the map $\omega \mapsto \varrho^{(\omega)}$ from the set Ω to the dual space $\mathcal{U}^*$:

- Translation invariance. Using definitions (4.39) and (6.1), we assume that

$$\varrho^{(\chi_x^{(\Omega)}(\omega))} = \varrho^{(\omega)} \circ \chi_x , \qquad x \in \mathfrak{L} = \mathbb{Z}^d . \tag{6.9}$$

- Measurability. The map $\omega \mapsto \varrho^{(\omega)}$ is measurable w.r.t. to the σ–algebra $\mathfrak{A}_\Omega$ on Ω and the Borel σ–algebra $\mathfrak{A}_{\mathcal{U}^*}$ of $\mathcal{U}^*$ generated by the weak*–topology. Note that a similar assumption is also used to define equilibrium for classical systems in disordered media, see, e.g., [Bo].

A map satisfying such properties is named here *a random invariant state* [BP6, Definition 3.1]. Such maps always exist in the one–dimension case if the norm $\|\Psi^{\mathrm{IP}}\|_{\mathcal{W}}$ of the interparticle interaction is finite. The same is true in any dimension if the inverse temperature $\beta \in \mathbb{R}^+$ is small enough. This is a consequence of the uniqueness of KMS, which is implied by the mentioned conditions. By using methods of constructive quantum field theory, one can also verify the existence of such random invariant passive states $\varrho^{(\omega)}$, $\omega \in \Omega$, at arbitrary dimension and any fixed $\beta \in \mathbb{R}^+$, if the interparticle interaction $\|\Psi^{\mathrm{IP}}\|_{\mathcal{W}}$ is small enough and (6.3) holds. See, for instance, [FU, Theorem 2.1] (together with [PW, Theorem 1.4]) for the small β case in quantum spin systems. See also [BP6, Sect. 3.3] for further discussions on this topic.

6.5 Macroscopic Paramagnetic Conductivity

For any short–range interaction $\Psi^{\mathrm{IP}} \in \mathcal{W}$, the limit

$$\Xi_{\mathrm{p}}(t) \doteq \lim_{l \to \infty} \mathbb{E}\left[\varrho^{(\omega)}(C_{\mathrm{p},l}^{(\omega)}(t))\right] \in \mathcal{B}(\mathbb{R}^d) \tag{6.10}$$

exists and is uniform for t on compacta. To see this, use the usual Lieb–Robinson bounds (Theorem 4.8 (iv)) to estimate (6.6) in the limit $l \to \infty$. Here, for any measurable $\mathbf{D}^{(\omega)} \in \mathcal{B}(\mathbb{R}^d)$, the expectation value $\mathbb{E}[\mathbf{D}^{(\omega)}] \in \mathcal{B}(\mathbb{R}^d)$ (associated with $\mathfrak{a}_\Omega$) is defined, w.r.t. the canonical orthonormal basis $\{e_q\}_{q=1}^d$ of $\mathbb{R}^d$, by the matrix entries

$$\{\mathbb{E}[\mathbf{D}^{(\omega)}]\}_{k,q} \doteq \mathbb{E}[\{\mathbf{D}\}_{k,q}] , \qquad k, q \in \{1, \ldots, d\} .$$

The function $\Xi_p \in C^1(\mathbb{R}; \mathcal{B}(\mathbb{R}^d))$ can be directly related to a linear response current, as suggested by (6.7)–(6.8). See [BP6, Theorem 4.2 (p)] for more details. [If one does not take expectation values of currents, one can also show that the limit $l \to \infty$ of $\varrho^{(\omega)}(\mathbf{J}_{p,l}^{(\omega)})$ almost everywhere exists and equals the expectation value, in the same limit, by using the Akcoglu–Krengel ergodic theorem, see [BPH3, BP6].]

[BP6, Theorem 7.1] asserts that

$$\Xi_p \in C^2(\mathbb{R}; \mathcal{B}(\mathbb{R}^d))$$

if $\Psi^{IP} \in \mathcal{W}$ and (4.56) holds with $\varsigma > 2d$. Now, we give a stronger version of this result which is an application of Lieb–Robinson bounds for multi–commutators (Theorems 4.10–4.11) of high orders. This new result on the regularity of the function Ξ_p of time has important consequences on the asymptotics of AC–Conductivity measures at high frequencies, see Theorem 6.5.

Theorem 6.1 (Regularity of the paramagnetic conductivity)
Let $\lambda \in \mathbb{R}_0^+$ and assume that the map $\omega \mapsto \varrho^{(\omega)}$ is a random invariant passive state and $\Psi^{IP} \in \mathcal{W}$ satisfies (6.3).
(i) Polynomial decay: Assume Ψ^{IP} satisfies (4.56). Then, for any $m \in \mathbb{N}$ satisfying $d(m+1) < \varsigma$, $\Xi_p \in C^{m+1}(\mathbb{R}; \mathcal{B}(\mathbb{R}^d))$ and, uniformly for t on compacta,

$$\partial_t^{m+1} \Xi_p(t) = \lim_{l \to \infty} \partial_t^{m+1} \mathbb{E}\left[\varrho^{(\omega)}(C_{p,l}^{(\omega)}(t)) \right] . \tag{6.11}$$

(ii) Exponential decay: Assume Ψ^{IP} satisfies (4.57). Then, for all $m \in \mathbb{N}$, $\Xi_p \in C^\infty(\mathbb{R}; \mathcal{B}(\mathbb{R}^d))$ and (6.11) holds true with the limit being uniform for t on compacta.

Remark 6.2 (Fermion systems with random Laplacians)
The same assertion holds for the random models treated in [BP6], i.e., for fermions on the lattice with short–range and translation invariant (cf. (6.3)) interaction $\Psi^{IP} \in \mathcal{W}$, random potentials (cf. (6.2)) and, additionally, random next neighbor hopping amplitudes. [So, Δ_d is replaced in [BP6] with a random Laplacian $\Delta_{\omega,\vartheta}$.] Similar to what is done here, disorder is defined in [BP6] via ergodic distributions of random potentials and hopping amplitudes.

The proof of this statement is a consequence of the following general lemma:

Lemma 6.3 *Let $\Psi \in \mathcal{W}$ and V be any potential such that*

$$\sup_{x \in \mathfrak{L}} \left\| V_{\{x\}} \right\|_{\mathcal{U}} < \infty . \tag{6.12}$$

Take $T \in \mathbb{R}_0^+$ and $B_0, B_1 \in \mathcal{U}_0$.
(i) Polynomial decay: Assume (4.56). Then, for any $m \in \mathbb{N}$ satisfying $dm < \varsigma$, $\mathcal{U}_0 \subseteq$ Dom(δ^m). Moreover, if $d(m+1) < \varsigma$,

$$\sum_{y\in\mathfrak{L}} \sup_{t\in[-T,T]} \sup_{x\in\mathfrak{L}} \left\| \left[\tau_t \circ \chi_x(B_1), \delta^m \circ \chi_y(B_0) \right) \right] \right\|_{\mathcal{U}} < \infty . \tag{6.13}$$

(ii) *Exponential decay: Assume (4.57). Then,*

$$\mathcal{U}_0 \subseteq \bigcap_{m\in\mathbb{N}} \mathrm{Dom}\left(\delta^m\right) \subset \mathcal{U}$$

and (6.13) holds true for all $m \in \mathbb{N}$.

Proof (i) Because of (6.12), assume w.l.o.g. that $\mathbf{V} = 0$. Take $t \in \mathbb{R}$, $n_0, n_1 \in \mathbb{N}$ and local elements $B_0 \in \mathcal{U}_{\Lambda_{n_0}}$ and $B_1 \in \mathcal{U}_{\Lambda_{n_1}}$. Then, we infer from Theorem 4.8 (ii) and (4.77)–(4.78) that, for any $x, y \in \mathfrak{L}$ and $n \in \mathbb{N}$,

$$\left\| \left[\tau_t \circ \chi_x(B_1), \delta^n \circ \chi_y(B_0) \right) \right] \right\|_{\mathcal{U}}$$
$$\leq \sum_{x_n\in\mathfrak{L}} \sum_{m_n\in\mathbb{N}_0} \sum_{\mathcal{Z}_n\in\mathcal{D}(x_n,m_n)} \cdots \sum_{x_1\in\mathfrak{L}} \sum_{m_1\in\mathbb{N}_0} \sum_{\mathcal{Z}_1\in\mathcal{D}(x_1,m_1)} \tag{6.14}$$
$$\left\| \left[\tau_t \circ \chi_x(B_1), \Psi_{\mathcal{Z}_n}, \ldots, \Psi_{\mathcal{Z}_1}, \chi_y(B_0) \right]^{(n+2)} \right\|_{\mathcal{U}} .$$

Therefore, we can directly use Lieb–Robinson bounds for multi–commutators of order $n + 2$ to bound (6.14): We combine Theorems 4.10 and 4.11 (i) with Eq. (4.116) to deduce from (6.14) that, for any $x, y \in \mathfrak{L}$ and $n \in \mathbb{N}$,

$$\left\| \left[\tau_t \circ \chi_x(B_1), \delta^n \circ \chi_y(B_0) \right) \right] \right\|_{\mathcal{U}}$$
$$\leq 2^{n+1} d^{\frac{\varsigma(n+1)}{2}} (1 + n_0)^\varsigma \| B_1 \|_{\mathcal{U}} \| B_0 \|_{\mathcal{U}} \tag{6.15}$$
$$\times \left(2\|\Psi\|_{\mathcal{W}} |t| e^{4\mathbf{D}|t|\|\Psi\|_{\mathcal{W}}} \left\| \mathbf{u}_{\cdot,n_1} \right\|_{\ell^1(\mathbb{N})} + (1 + n_1)^\varsigma \right)$$
$$\times \left(\sup_{x\in\mathfrak{L}} \left(\sum_{m\in\mathbb{N}_0} (1 + m)^\varsigma \sum_{\mathcal{Z}\in\mathcal{D}(x,m)} \|\Psi_{\mathcal{Z}}\|_{\mathcal{U}} \right) \right)^n$$
$$\times \sum_{x_n\in\mathfrak{L}} \cdots \sum_{x_1\in\mathfrak{L}} \left(\sum_{T\in\mathcal{T}_{n+2}} \prod_{\{j,l\}\in T} \frac{1}{(1 + |x_j - x_l|)^{\varsigma(\max\{\mathfrak{d}_T(j),\mathfrak{d}_T(l)\})^{-1}}} \right)$$

with $x_0 \doteq y \in \mathfrak{L}$ and $x_{n+1} \doteq x \in \mathfrak{L}$. If $\Psi \in \mathcal{W}$ and Condition (4.56) holds true, then one easily verifies (4.80) with $\mathbf{v}_m = (1 + m)^\varsigma$. Recall also that the condition $\varsigma > (n + 1) d$ yields (4.84) with $k = n + 1$. Using these observations, one directly arrives at (6.13), starting from (6.15).

Remark that $\mathcal{U}_0 \subseteq \mathrm{Dom}(\delta^n)$ is proven exactly in the same way. In fact, it is easier to prove and only requires the condition $\varsigma > nd$ because we have in this case multi–commutators of only order $n + 1$.

(ii) The proof is very similar to the polynomial case. We omit the details. See Theorem

4.11 (ii) and (4.66), and in the case (4.57) holds and $\Psi \in \mathcal{W}$, note again that Condition (4.80) is satisfied with $\mathbf{v}_m = e^{m\varsigma}$. ∎

We are now in position to prove Theorem 6.1.

Proof Fix $k, q \in \{1, \ldots, d\}, t \in \mathbb{R}$ and $m \in \mathbb{N}$. By Theorem 4.8 (i), $\tau^{(\omega)} \doteq \{\tau_t^{(\omega)}\}_{t \in \mathbb{R}}$ is a C_0–group of $*$–automorphisms with generator $\delta^{(\omega)}$. It is, indeed, associated with the interaction (6.4) and the potential defined by (6.2). If Ψ^{IP} satisfies (4.56), then Condition (4.56) also holds true for the full interaction (6.4). A similar observation can be made when Ψ^{IP} satisfies (4.57).

Paramagnetic current observables (6.5) are local elements, i.e., $I_{(x,y)} \in \mathcal{U}_0$ for any $(x, y) \in \mathfrak{L}^2$. Then, by Lemma 6.3, we thus compute from (6.6) that, for any $m \in \mathbb{N}$ such that $\mathcal{U}_0 \subseteq \mathrm{Dom}(\delta^m)$,

$$\partial_t^{m+1} \left\{ \mathbb{E}\left[\varrho^{(\omega)}(C_{\mathrm{p},l}^{(\omega)}(t)) \right] \right\}_{k,q} \tag{6.16}$$
$$= -\frac{1}{|\Lambda_l|} \sum_{x,y \in \Lambda_l} \mathbb{E}\left[\varrho^{(\omega)} \left(i[\tau_{-t}^{(\omega)} \circ (\delta^{(\omega)})^m (I_{(y+e_q,y)}), I_{(x+e_k,x)}] \right) \right] .$$

The last function of $\omega \in \Omega$ in the expectation value $\mathbb{E}[\ \cdot\ ]$ (associated with $\mathfrak{a}_\Omega$) is measurable, because $\omega \mapsto \varrho^{(\omega)}$ is, by definition, a random invariant state while one can check that the map

$$\omega \mapsto i[\tau_{-t}^{(\omega)} \circ (\delta^{(\omega)})^m (I_{(y+e_q,y)}), I_{(x+e_k,x)}]$$

from Ω to $\mathcal{U}$ is continuous, using Theorem 4.8 and the second Trotter–Kato approximation theorem [EN, Chap. III, Sect. 4.9]. Additionally, if $\varrho^{(\omega)}$ is a passive state w.r.t. to $\tau^{(\omega)}$ for any $\omega \in \Omega$ then $\varrho^{(\omega)} = \varrho^{(\omega)} \circ \tau_t^{(\omega)}$, see [PW, Theorem 1.1]. Therefore, it follows from (6.16) that

$$\partial_t^{m+1} \left\{ \bar{\varrho} \left(C_{\mathrm{p},l}^{(\omega)}(t) \right) \right\}_{k,q} \tag{6.17}$$
$$= \frac{1}{|\Lambda_l|} \sum_{x,y \in \Lambda_l} \mathbb{E}\left[\varrho^{(\omega)} \left(i[\tau_t^{(\omega)} \left(I_{(x+e_k,x)} \right), (\delta^{(\omega)})^m (I_{(y+e_q,y)})] \right) \right] .$$

Now, if (6.3) and (6.9) hold true, then, by using the fact that $\mathfrak{a}_\Omega$ is also a translation invariant probability measure (it is even ergodic), we obtain from (6.17) that, for any $m \in \mathbb{N}$ such that $\mathcal{U}_0 \subseteq \mathrm{Dom}(\delta^m)$,

$$\partial_t^{m+1} \left\{ \bar{\varrho} \left(C_{\mathrm{p},l}^{(\omega)}(t) \right) \right\}_{k,q} \tag{6.18}$$
$$= \sum_{y \in \mathfrak{L}} \xi_l(y) \mathbb{E}\left[\varrho^{(\omega)} \left(i[\tau_t^{(\omega)} \left(I_{(e_k,0)} \right), (\delta^{(\omega)})^m \circ \chi_y(I_{(e_q,0)})] \right) \right]$$

with

$$\xi_l(y) \doteq \frac{1}{|\Lambda_l|} \sum_{x \in \Lambda_l} 1_{\{y \in \Lambda_l - x\}} \in [0, 1] , \qquad y \in \mathfrak{L}, \, l \in \mathbb{R}^+ .$$

For any $l \in \mathbb{R}^+$, the map $y \mapsto \xi_l(y)$ on $\mathfrak{L}$ has finite support and, for any $y \in \mathfrak{L}$,

$$\lim_{l \to \infty} \xi_l(y) = 1 . \tag{6.19}$$

As a consequence, if (i) Ψ^{IP} satisfies (4.56) and $d(m + 1) < \varsigma$ or (ii) Ψ^{IP} satisfies (4.57), then, by combining Lemma 6.3 with Lebesgue's dominated convergence theorem, one gets from (6.10) and (6.18)–(6.19) that the map

$$t \mapsto \partial_t^{m+1} \left\{ \mathbb{E} \left[\varrho^{(\omega)}(C_{\mathrm{p},l}^{(\omega)}(t)) \right] \right\} = \mathbb{E} \left[\partial_t^{m+1} \varrho^{(\omega)}(C_{\mathrm{p},l}^{(\omega)}(t)) \right]$$

converges uniformly on compacta, as $l \to \infty$, to the continuous function $\partial_t^{m+1} \Xi_{\mathrm{p}} \in C(\mathbb{R}; \mathcal{B}(\mathbb{R}^d))$. ∎

6.6 AC–Conductivity Measure

By applying [BP6, Theorems 5.2 and 5.6 (p), Remark 5.3] to the interacting fermion system under consideration we get a *Lévy–Khintchine representation* of the paramagnetic (in–phase) conductivity Ξ_{p}: Assume Ψ^{IP} satisfies (4.56) with $\varsigma > 2d$ (polynomial decay). Then, there is a unique finite and symmetric $\mathcal{B}_+(\mathbb{R}^d)$–valued measure μ on $\mathbb{R}$ such that, for any $t \in \mathbb{R}$,

$$\Xi_{\mathrm{p}}(t) = -\frac{t^2}{2} \cdot (\{0\}) + \int_{\mathbb{R} \setminus \{0\}} (\cos(t\nu) - 1) \, \nu^{-2} \mu(d\nu) . \tag{6.20}$$

Here, $\mathcal{B}_+(\mathbb{R}^d) \subset \mathcal{B}(\mathbb{R}^d)$ stands for the set of positive linear operators on $\mathbb{R}^d$, i.e., symmetric operators w.r.t. to the canonical scalar product of $\mathbb{R}^d$ with positive eigenvalues. The (in–phase) *AC–conductivity measure* is defined from the measure μ as follows:

Definition 6.4 (*AC–conductivity measure*) We name the Lévy measure μ_{AC}, the restriction of $\nu^{-2} \mu(d\nu)$ to $\mathbb{R} \setminus \{0\}$, the (in–phase) AC–conductivity measure.

Indeed, by [BP6, Theorems 5.1 and 5.6 (p)], one checks that μ_{AC} quantifies the energy (or heat) production Q per unit volume due to the component of frequency $\nu \in \mathbb{R} \setminus \{0\}$ of the electric field, in accordance with Joule's law in the AC–regime: Indeed, for any smooth electric field $E(t) = \mathcal{E}(t) \vec{w}$ with $\vec{w} \in \mathbb{R}^d, \mathcal{E} \doteq -\partial_t \mathcal{A}(t)$ and $\mathcal{A} \in C_0^\infty(\mathbb{R}; \mathbb{R})$, the total heat per unit volume produced by the electric field (after being switch off) is equal to

$$Q = \frac{1}{2} \int_{\mathbb{R}} ds_1 \int_{\mathbb{R}} ds_2 \mathcal{E}_{s_2} \mathcal{E}_{s_1} \langle \vec{w}, \Xi_p (s_1 - s_2) \vec{w} \rangle_{\mathbb{R}^d} .$$

If the Fourier transform $\hat{\mathcal{E}}$ of $\mathcal{E} \in C_0^\infty (\mathbb{R}; \mathbb{R})$ has support away from $\nu = 0$, then

$$Q = \frac{1}{2} \int_{\mathbb{R} \setminus \{0\}} |\hat{\mathcal{E}} (\nu)|^2 \langle \vec{w}, \mu_{AC} (d\nu) \vec{w} \rangle_{\mathbb{R}^d} .$$

Moreover, by using [BP6, Theorems 4.2 and 5.6 (p)] together with simple computations, one checks that the in–phase linear response currents J_{in}, which is the component of the total current producing heat, also called active current, is equal in this case to

$$J_{in} (t) = \int_{\mathbb{R} \setminus \{0\}} \hat{\mathcal{E}} (\nu) \, e^{i\nu t} \, \mu_{AC} (d\nu) \, \vec{w} .$$

By (6.20) and Definition 6.4, observe that the AC–conductivity measure μ_{AC} of the system under consideration is a Lévy measure. This is reminiscent of experimental observations of other quantum phenomena like (subrecoil) laser cooling [BBAC]. In fact, an alternative effective description of the phenomenon of linear conductivity by using Lévy processes in Fourier space is discussed in [BP6, Sect. 6].

The explicit form of the conductivity measure for concrete models (like the Hubbard model, for instance) is still an open problem. However, in [BP6, Sect. 5.3], we were able to qualitatively compare the AC–conductivity measure associated with the celebrated Drude model with the Lévy measure μ_{AC} given by Definition 6.4. Indeed, the (in–phase) AC–conductivity measure obtained from the Drude model is absolutely continuous w.r.t. the Lebesgue measure with the function

$$\nu \mapsto \vartheta_T (\nu) \sim \frac{T}{1 + T^2 \nu^2} \tag{6.21}$$

being the corresponding Radon–Nikodym derivative. Here, the *relaxation time* $T > 0$ is related to the mean time interval between two collisions of a charged carrier with defects in the crystal. See for instance [BPH4, Sect. 1] for more discussions. This measure *heavily overestimates* μ_{AC} at high frequencies. Indeed, as explained in [BP6, Sect. 5.3], by finiteness of the positive measure μ, the AC–conductivity measure satisfies

$$\mu_{AC} ([\nu, \infty)) \leq \nu^{-2} \mu ([\nu, \infty)) \leq \nu^{-2} \mu (\mathbb{R}) , \qquad \nu \in \mathbb{R}^+ , \tag{6.22}$$

provided Ψ^{IP} satisfies (4.56) with $\varsigma > 2d$. The same property of course holds for negative frequencies, by symmetry of μ (w.r.t. ν). Compare (6.22) with (6.21). From Theorem 6.1, much stronger results on the frequency decay of μ_{AC} can be obtained if the interaction Ψ^{IP} is fast decaying in space:

Theorem 6.5 (Moments of AC–conductivity measures)

Let $\lambda \in \mathbb{R}_0^+$, $\Psi^{\mathrm{IP}} \in \mathcal{W}$ satisfying (6.3), and assume that the map $\omega \mapsto \varrho^{(\omega)}$ is a random invariant passive state.

(i) *Polynomial decay:* Assume Ψ^{IP} satisfies (4.56) with $\varsigma > 2d$. Then, for any $m \in \mathbb{N}$ satisfying $d(m + 1) < \varsigma$,

$$\int_{\mathbb{R}\backslash\{0\}} \nu^{m+1} \mu_{\mathrm{AC}} (\mathrm{d}\nu) \in \mathcal{B}_+(\mathbb{R}^d) , \tag{6.23}$$

i.e., the $(m + 1)$-th moment of the measure μ_{AC} exists.

(ii) *Exponential decay:* Assume Ψ^{IP} satisfies (4.57). Then, (6.23) holds true for all $m \in \mathbb{N}$.

Proof By (6.20) and Lebesgue's dominated convergence theorem, for any $t \in \mathbb{R}$,

$$\partial_t^2 \Xi_{\mathrm{p}} (t) = - \int_{\mathbb{R}} \cos (t\nu) \mu (\mathrm{d}\nu) = - \int_{\mathbb{R}} \mathrm{e}^{it\nu} \mu (\mathrm{d}\nu) ,$$

provided $\varsigma > 2d$ in (4.56) (with $\Psi = \Psi^{\mathrm{IP}}$). In other words, the finite and symmetric $\mathcal{B}_+(\mathbb{R}^d)$–valued measure μ on $\mathbb{R}$ can be seen as the Fourier transform of $-\partial_t^2 \Xi_{\mathrm{p}} (t)$ or, that is, as the characteristic function of μ. Therefore, by well–known properties of characteristic functions (see, e.g., [D, Theorem 3.3.9.] for the special case $n = 2$ and [Kl, Theorem 15.34] for the general case $n \in 2\mathbb{N}_0$), for any even $n \in \mathbb{N}_0$, $\partial_t^2 \Xi_{\mathrm{p}} \in C^n(\mathbb{R}; \mathcal{B}(\mathbb{R}^d))$ implies that

$$\int_{\mathbb{R}} \nu^n \mu (\mathrm{d}\nu) \in \mathcal{B}_+(\mathbb{R}^d) .$$

If $m \in \mathbb{N}_0$ is odd, then, by the above assertion for $n < m$ and the symmetry of the measure μ (which follows from the symmetry of μ_{AC}), we conclude that

$$\int_{\mathbb{R}} \nu^m \mu (\mathrm{d}\nu) = 0 \in \mathcal{B}_+(\mathbb{R}^d) .$$

This observation combined with Theorem 6.1 and Definition 6.4 yields Assertions (i)–(ii). $\qed$

Remark 6.6 (*Fermion systems with random Laplacians*)
The same assertion holds for the random models treated in [BP6]. See also Remark 6.2.

This last theorem is a significant improvement of the asymptotics (6.22) of [BP6] and is a straightforward application of Lieb–Robinson bounds for multi–commutators of high orders (Theorems 4.10–4.11), see Lemma 6.3.

Acknowledgements This research is supported by the agency FAPESP under Grant 2013/13215-5 as well as by the Basque Government through the grant IT641-13 and the BERC 2014-2017 program and by the Spanish Ministry of Economy and Competitiveness MINECO: BCAM Severo Ochoa accreditation SEV-2013-0323, MTM2014-53850.

References

[AKLT]   I. Affleck, T. Kennedy, E.H. Lieb, H. Tasaki, Valence bond ground states in isotropic quantum antiferromagnets. Commun. Math. Phys. **115**(3), 477–528 (1988)

[AM]   H. Araki, H. Moriya, Equilibrium statistical mechanics of Fermion lattice system. Rev. Math. Phys. **15**, 93–198 (2003)

[AT]   P. Acquistapace, B. Terreni, A unified approach to abstract linear nonautonomous parabolic equations. Rend. Sem. Mat. Univ. Padova **78**, 47–107 (1987)

[B]   N. Bohr, *Physique atomique et connaissance humaine*, translation by ed. E. Bauer, R. Omnès, C. Chevalley (Paris, Editions Gallimard, 1991)

[BB]   V. Bach, J.-B. Bru, Diagonalizing quadratic Bosonic operators by non-autonomous flow equation. Mem. AMS **240**, 1138 (2016). http://dx.doi.org/10.1090/memo/1138

[BMNS]   S. Bachmann, S. Michalakis, B. Nachtergaele, R. Sims, Automorphic equivalence within gapped phases of quantum lattice systems. Commun. Math. Phys. **309**, 835–871 (2012)

[BBAC]   F. Bardou, J.-P. Bouchaud, A. Aspect, C. Cohen-Tannoudji, *Lévy Statistics and Laser Cooling* (Cambridge University Press, Cambridge, 2001). (Cambridge Books Online)

[BC]   M.H. Brynildsen, H.D. Cornean, On the Verdet constant and Faraday rotation for graphene-like materials. Rev. Math. Phys. **25**(4), 1350007-1–28 (2013)

[Bo]   A. Bovier, *Statistical Mechanics of Disordered Systems: A Mathematical Perspective*, Cambridge Series in Statistical and Probabilistic Mathematics (Cambridge University Press, Cambridge, 2006)

[BR1]   O. Bratteli, D.W. Robinson, *Operator Algebras and Quantum Statistical Mechanics*, vol. I, 2nd edn. (Springer, New York, 1996)

[BR2]   O. Bratteli, D.W. Robinson, *Operator Algebras and Quantum Statistical Mechanics*, vol. II, 2nd edn. (Springer, New York, 1996)

[BP1]   J.-B. Bru, W. de Siqueira Pedra, Effect of a locally repulsive interaction on s-wave superconductors. Rev. Math. Phys. **22**(3), 233–303 (2010)

[BP2]   J.-B. Bru, W. de Siqueira Pedra, Non-cooperative equilibria of Fermi systems with long range interactions. Mem. AMS **224**(1052) (2013)

[BP3]   J.-B. Bru, W. de Siqueira Pedra, Inhomogeneous Fermi and quantum spin systems on lattices. J. Math. Phys. **53**, 123301-1–25 (2012)

[BP4]   J.-B. Bru, W. de Siqueira Pedra, Microscopic foundations of Ohm and Joule's Laws – The relevance of thermodynamics. Mathematical results in quantum mechanics, in *Proceedings of the QMath12 Conference*, eds. by P. Exner, W. König, H. Neidhardt (World Scientific, Singapore, 2014)

© The Author(s) 2017

J.-B. Bru and W. de Siqueira Pedra, *Lieb-Robinson Bounds for Multi-commutators and Applications to Response Theory*, SpringerBriefs in Mathematical Physics, DOI 10.1007/978-3-319-45784-0

[BP5]   J.-B. Bru, W. de Siqueira Pedra, Microscopic conductivity of lattice fermions at
          equilibrium - Part II: interacting particles. Lett. Math. Phys. **106**(1), 81–107 (2016)
[BP6]   J.-B. Bru, W. de Siqueira Pedra, From the 2nd law of thermodynamics to the AC-
          conductivity measure of interacting fermions in disordered media. M3AS: Math.
          Models Methods Appl. Sci. **25**(14), 2587–2632 (2015)
[BPH1]   J.-B. Bru, W. de Siqueira Pedra, C. Hertling, Heat production of non-interacting
          fermions subjected to electric fields. Commun. Pure Appl. Math. **68**(6), 964–1013
          (2015)
[BPH2]   J.-B. Bru, W. de Siqueira Pedra, C. Hertling, Microscopic conductivity of lattice
          fermions at equilibrium - Part I: non-interacting particles. J. Math. Phys. **56**, 051901-
          1–51 (2015)
[BPH3]   J.-B. Bru, W. de Siqueira Pedra, C. Hertling, AC-conductivity measure from heat
          production of free fermions in disordered media. Arch. Ration. Mech. Anal. **220**(2),
          445–504 (2016). doi:10.1007/s00205-015-0935-1
[BPH4]   J.-B. Bru, W. de Siqueira Pedra, C. Hertling, Macroscopic conductivity of free fermi-
          ons in disordered media. Rev. Math. Phys. **26**(5), 1450008-1–25 (2014)
[BvN]   G. Birkhoff, J. Von Neumann, The logic of quantum mechanics. Ann. Math. **37**(4),
          823–843 (1936)
[C]   O. Caps, *Evolution Equations in Scales of Banach Spaces* (B.G, Teubner, Stuttgart-
          Leipzig-Wiesbaden, 2002)
[Ch]   M. Cheneau et al., Light-cone-like spreading of correlations in a quantum many-body
          system. Nature **481**, 484–487 (2012)
[DG]   N. Dombrowski, F. Germinet, Linear response theory for random Schrodinger oper-
          ators and noncommutative integration. Markov. Proc. Rel. F. **18**, 403–426 (2008)
[D]   R. Durrett, *Probability: Theory and Examples*, 4th edn. (Cambridge University Press,
          Cambridge, 2010)
[Dix]   J. Dixmier, *Les C*-algèbres et leurs représentations* (Gauthier-Villars, 1969)
[E]   G. Emch, *Algebraic Methods in Statistical Mechanics and Quantum Field Theory*
          (Willey-Interscience, New York, 1972)
[EN]   K.-J. Engel, R. Nagel, *One-Parameter Semigroups for Linear Evolution Equations*
          (Springer, New York, 2000)
[F]   C. Flori, *A First Course in Topos Quantum Theory*, vol. 868, Lecture Notes in Physics
          (Springer, Berlin, 2013)
[FK]   S. French, D. Krause, *Identity in Physics: A Historical, Philosophical, and Formal
          Analysis* (Oxford University Press, New York, 2006)
[FMU]   J. Fröhlich, M. Merkli, D. Ueltschi, Dissipative transport: thermal contacts and tun-
          nelling junctions. Ann. Henri Poincaré **4**, 897–945 (2003)
[FU]   J. Fröhlich, D. Ueltschi, Some properties of correlations of quantum lattice systems
          in thermal equilibrium. J. Math. Phys. **56**, 053302-1–14 (2015)
[G]   M.D. Girardeau, Permutation symmetry of many-particle wave functions. Phys. Rev.
          139, B500–B508 (1965)
[H1]   R. Haag, The mathematical structure of the Bardeen–Cooper–Schrieffer Model. Il
          Nuovo Cimento. **XXV**(2), 287–299 (1962)
[H2]   R. Haag, Some people and some problems met in half a century of commitment to
          mathematical physics. Eur. Phys. J. H **35**, 263–307 (2010)
[H]   L. Hörmander, *The Analysis of Linear Partial Differential Operators*, vol. 1–4
          (Springer, Berlin, 1983–1984)
[I]   R.B. Israel, *Convexity in the Theory of Lattice Gases* (Princeton University Press,
          Princeton, 1979)
[KR1]   R.V. Kadison, J.R. Ringrose, *Fundamentals of the Theory of Operator Algebras,
          Volume I: Elementary Theory*, vol. 15, Graduate Studies in Mathematics, Reprinted
          by the AMS (1997)
[KR2]   R.V. Kadison, J.R. Ringrose, *Fundamentals of the Theory of Operator Algebras,
          Volume II: Advanced Theory*, vol. 15, Graduate Studies in Mathematics, Reprinted
          by the AMS (1997)

[K1]   T. Kato, Integration of the equation of evolution in a Banach space. J. Math. Soc. Jpn. **5**, 208–234 (1953)

[K2]   T. Kato, Linear evolution equations of "hyperbolic" type. J. Fac. Sci. Univ. Tokyo **17**, 241–258 (1970)

[K3]   T. Kato, Linear evolution equations of 'hyperbolic' type II. J. Math. Soc. Jpn. **25**, 648–666 (1973)

[K4]   T. Kato, Abstract evolution equations, linear and quasilinear, revisited, in *Functional Analysis and Related Topics*, vol. 1540, Lecture Notes in Mathematics, ed. by H. Komatsu (Springer, New York, 1993), pp. 103–125

[KLM]   A. Klein, O. Lenoble, P. Müller, On Mott's formula for the ac-conductivity in the Anderson model. Ann. Math. **166**, 549–577 (2007)

[KM1]   A. Klein, P. Müller, The conductivity measure for the Anderson model. J. Math. Phys. Anal. Geom. **4**, 128–150 (2008)

[KM2]   A. Klein, P. Müller, AC-conductivity and electromagnetic energy absorption for the Anderson model in linear response theory. Markov Process. Relat. Fields **21**, 575–590 (2015)

[Kl]   A. Klenke, *Probability Theory - A Comprehensive Course* (Springer, London, 2008)

[LM]   J.M. Leinaas, J. Myrheim, On the theory of identical particles. IL Nuovo Cimento B **37**(1), 1–23 (1977)

[LTW]   A. Lagendijk, B. van Tiggelen, D.S. Wiersma, Fifty years of Anderson localization. Phys. Today **62**(8), 24–29 (2009)

[LR]   E.H. Lieb, D.W. Robinson, The finite group velocity of quantum spin systems. Commun. Math. Phys. **28**, 251–257 (1972)

[LY1]   E.H. Lieb, J. Yngvason, The physics and mathematics of the second law of thermodynamics. Phys. Rep. **310**, 1–96 (1999)

[LY2]   E.H. Lieb, J. Yngvason, The mathematical structure of the second law of thermodynamics. Curr. Dev. Math. **2001**, 89–129 (2002)

[N]   B. Nachtergaele, Quantum spin systems, *Article for the Encyclopedia of Mathematical Physics* (Elsevier). arXiv:math-ph/0409006

[Na]   M.A. Naimark, Rings with involutions, *Uspehi Matematiceskih Nauk (N.S.)*, **3**, 52–145 (1948) (in Russian)

[Ne]   G. Nenciu, Dynamics of band electrons in electric and magnetic fields: rigorous justification of the effective Hamiltonians. Rev. Mod. Phys. **63**(1), 91–127 (1991)

[NOS]   B. Nachtergaele, Y. Ogata, R. Sims, Propagation of correlations in quantum lattice systems. J. Stat. Phys. **124**(1), 1–13 (2006)

[NS]   B. Nachtergaele, R. Sims, Lieb-Robinson bounds in quantum many-body physics. Contemp. Math. **529**, 141–176 (2010)

[NS1]   S.R. Nagel, S.E. Schnatterly, Frequency dependence of the Drude relaxation time in metal films. Phys. Rev. B **9**(4), 1299–1303 (1974)

[NS2]   S.R. Nagel, S.E. Schnatterly, Frequency dependence of the Drude relaxation time in metal films: further evidence for a two-carrier model. Phys. Rev. B **12**(12), 6002–6005 (1975)

[NZ]   V.A. Zagrebnov, H. Neidhardt, Linear non-autonomous Cauchy problems and evolution semigroups. Adv. Differ. Equ. **14**, 289–340 (2009)

[P]   A. Pazy, *Semigroups of Linear Operators and Applications to Partial Differential Equation*, vol. 44, Applied Mathematical Sciences (Springer, New York, 1983)

[PW]   W. Pusz, S.L. Woronowicz, Passive states and KMS states for general quantum systems. Commun. Math. Phys. **58**, 273–290 (1978)

[R]   J. Renn, in *Schrödinger and the Genesis of Wave Mechanics*, ESI Lectures in Mathematics and Physics, ed. by W.L. Reiter, J. Yngvason (EMS, 2013), pp. 9–36

[Ro]   H. Robbins, A remark on Stirling's formula. Am. Math. Monthly **62**, 26–29 (1955)

[Ros]   A. Rosenberg, The number of irreducible representations of simple rings with no minimal ideals. Am. J. Math. **75**, 523–530 (1953)

[RS1]   M. Reed, B. Simon, *Methods of Modern Mathematical Physics, Vol. I: Functional Analysis* (Academic Press, New York, 1972)

[RS2]   M. Reed, B. Simon, *Methods of Modern Mathematical Physics, Vol. II: Fourier Analysis, Self-Adjointness* (Academic Press, New York, 1975)

[Ru]   W. Rudin, *Functional Analysis* (McGraw-Hill Science, New York, 1991)

[S]   R. Schnaubelt, Asymptotic behaviour of parabolic nonautonomous evolution equations, in *Functional Analytic Methods for Evolution Equations* ed. by M. Iannelli, R. Nagel, S. Piazzera (Springer, Heidelberg, 2004), pp. 401–472

[SE]   J.B. Smith, H. Ehrenreich, Frequency dependence of the optical relaxation time in metals. Phys. Rev. B **25**(2), 923–930 (1982)

[Si]   R. Sims, Lieb-Robinson bounds and quasi-locality for the dynamics of many-body quantum systems. Mathematical results, in quantum physics, ed. by P. Exner in *Proceedings of the QMath 11 Conference in Hradec Kralove, Czech Republic, 2010* (World Scientific, Hackensack, 2011), pp. 95–106

[Sh]   E. Shrödinger, What is matter? Sci. Am. **189**, 52–57 (1953)

[So]   E.H. Sondheimer, The mean free path of electrons in metals. Adv. Phys. **50**(6), 499–537 (2001)

[T]   M.-L. Thèye, Investigation of the optical properties of Au by means of thin semi-transparent films. Phys. Rev. B **2**, 3060 (1970)

[TW]   W. Thirring, A. Wehrl, On the mathematical structure of the B.C.S.-model. Commun. Math. Phys. **4**, 303–314 (1967)

[W]   I.R.F. Wagner, Ph.D. thesis: algebraic approach towards conductivity in ergodic media, München, University, Dissertation (2013)

[Wea]   N. Weaver, *Forcing for Mathematicians* (World Scientific Publishing, Singapore, 2014)

[Wi]   F. Wilczek, Quantum mechanics of fractional-spin particles. Phys. Rev. Lett. **49**(14), 957–959 (1982)

[YRMK]   S.J. Youn, T.H. Rho, B.I. Min, K.S. Kim, Extended Drude model analysis of noble metals. Phys. Stat. Sol. (b) **244**(49), 1354–1362 (2007)

Index

© The Author(s) 2017   107
J.-B. Bru and W. de Siqueira Pedra, *Lieb-Robinson Bounds for Multi-commutators
and Applications to Response Theory*, SpringerBriefs in Mathematical Physics,
DOI 10.1007/978-3-319-45784-0

Printed in the United States
By Bookmasters